EXPORTING THE BOMB

A volume in the series
Cornell Studies in Security Affairs
edited by Robert J. Art, Robert Jervis, and Stephen M. Walt

A list of titles in this series is available at
www.cornellpress.cornell.edu.

EXPORTING THE BOMB

Technology Transfer and
the Spread of Nuclear Weapons

Matthew Kroenig

CORNELL UNIVERSITY PRESS   ITHACA AND LONDON

Copyright © 2010 by Cornell University

All rights reserved. Except for brief quotations in a review, this book, or parts thereof, must not be reproduced in any form without permission in writing from the publisher. For information, address Cornell University Press, Sage House, 512 East State Street, Ithaca, New York 14850.

First published 2010 by Cornell University Press
First printing, Cornell Paperbacks, 2010
Printed in the United States of America

Library of Congress Cataloging-in-Publication Data

Kroenig, Matthew.
  Exporting the bomb : technology transfer and the spread of nuclear weapons / Matthew Kroenig.
    p. cm. — (Cornell studies in security affairs)
  Includes bibliographical references and index.
  ISBN 978-0-8014-4857-7 (cloth : alk. paper)
  ISBN 978-0-8014-7640-2 (pbk. : alk. paper)
  1. Nuclear weapons—Political aspects. 2. Arms transfers—Political aspects. 3. Technology transfer—Political aspects. 4. Military assistance—Political aspects. 5. Nuclear nonproliferation—Political aspects. 6. Security, International. I. Title. II. Series: Cornell studies in security affairs.
  JZ5665.K76 2010
  327.1'747—dc22   2009038046

Cornell University Press strives to use environmentally responsible suppliers and materials to the fullest extent possible in the publishing of its books. Such materials include vegetable-based, low-VOC inks and acid-free papers that are recycled, totally chlorine-free, or partly composed of nonwood fibers. For further information, visit our website at www.cornellpress.cornell.edu.

Cloth printing   10 9 8 7 6 5 4 3 2 1
Paperback printing   10 9 8 7 6 5 4 3 2 1

Contents

Acknowledgments	vii
List of Abbreviations and Acronyms	xi
Introduction: The Problem of Nuclear Assistance	1
1. Explaining Nuclear Assistance	10
2. The Correlates of Nuclear Assistance	50
3. Israel's Nuclear Program: French Assistance and U.S. Resistance	67
4. Common Enemies, Growling Dogs, and A. Q. Khan's Pakistan: Nuclear Supply in Other Countries	111
5. Importing the Bomb: Nuclear Assistance and Nuclear Proliferation	151
Conclusion: Preventing Nuclear Proliferation	173
Appendixes:	
A. Data Appendix for Chapter 2	191
B. Data Appendix for Chapter 5	195
C. Cases of Sensitive Nuclear Assistance	197
D. Selected Cases of Nonsensitive Nuclear Assistance	200
E. Selected Cases of Nonassistance	202
Bibliography	205
Index	227

Acknowledgments

This book traces its origins to work I performed as a military analyst in the U.S. government.[1] I was drafting a strategic assessment on how great powers would react to a particular nuclear proliferation issue and was surprised by my own analysis. We often hear that nuclear proliferation poses a general threat to international peace and security and that, for this reason, great powers can work together to combat the threat of nuclear proliferation. Yet, with the issue I was working on, I found great variety in the way countries approached the problem: some countries seemed very threatened by nuclear proliferation and were willing to do almost anything to stop it, others seemed less concerned, and still others were actually helping other countries develop their nuclear weapons programs. I could not understand why countries took such different approaches to what is often thought to be a common problem of nuclear proliferation. More important, I wanted to know why some countries were willing to go so far as to help other countries acquire the deadliest weapons known to mankind. This book is the result of that initial curiosity.

A variety of individuals and institutions played a significant role in the production of this book. I owe a special debt of gratitude to advisors in the Department

---

[1]. All statements of fact, opinion, or analysis expressed are those of the author and do not reflect the official positions or views of the CIA or any other U.S. government agency. Nothing in its contents should be construed as asserting or implying U.S. government authentication of information or Agency endorsement of the author's views. This material has been reviewed by the CIA to prevent the disclosure of classified information.

of Political Science at the University of California at Berkeley. Steve Weber taught me that an argument can be creative, provocative, and correct, all at the same time. I thank M. Steven Fish for his thoughtful guidance on navigating the academy and the wider world beyond it. I thank Giacomo Chiozza for his invaluable advice on measuring and quantitatively analyzing international political events. Michael Nacht provided deep knowledge on nuclear weapons and arms control issues, along with a neverending supply of anecdotes about the policymakers whom I was studying.

An early fellowship with the Institute on Global Conflict and Cooperation (IGCC) at the University of California convinced me that nuclear proliferation was a subject that demanded further scholarly inquiry. I thank Susan Shirk for her work in establishing the Public Policy and Nuclear Threats Program at the IGCC and for saving me from writing on a less important topic.

A research workshop sponsored by the Committee for the Analysis of Military Operations and Strategy helped me to refine my proposal. I thank Stephen Biddle, Elizabeth Kier, and Dan Reiter for their detailed comments on my prospectus.

I spent a year as a fellow at the Center for International Security and Cooperation at Stanford University. I thank Lynn Eden for welcoming me into the vibrant intellectual community that she has helped to create there. I am also deeply indebted to Scott Sagan, whose insights and expertise on the scholarly study of nuclear proliferation assisted me at every stage of this project, all the way from helping me to settle on a question at the beginning of my research to reading and commenting on the entire manuscript upon its completion.

A postdoctoral fellowship in the Managing the Atom Project and the International Security Program at the Belfer Center for Science and International Affairs at Harvard University provided me with the uninterrupted time and supportive environment that helped bring this project to fruition. I thank Steve Miller and Martin Malin for the opportunity to participate in their programs and to take advantage of Harvard's many resources.

I am also grateful for comments provided on this research project over the years by Graham Allison, Victor Asal, Naazneen Barma, Kyle Beardsley, Matthew Bunn, Alisa Carrigan, Ashton Carter, Jonathan Caverley, Alex Downes, Brent Durbin, James Fearon, Edward Fogarty, Matthew Fuhrmann, Erik Gartzke, Charles Glaser, Ron Hassner, Siegfried Hecker, John Holdren, Robert Jervis, Peter Katzenstein, Sarah Kreps, Sean Lynn-Jones, Alex Montgomery, Danielle McLaughlin, John Mearsheimer, Jonathan Monten, Patrick Morgan, John Mueller, Tatishe Nteta, Robert Pape, David Patel, Robert Powell, Ely Ratner, Phillip Roessler, Todd Sechser, Jacob Shapiro, Harold Smith, Regine Spector, Paul Stockton, Kevin Wallsten, Christopher Way, Stephen Walt, Dean Wilkening, and Zachary Zwald. My apologies to anyone I forgot.

I have also benefited from generous financial support. I thank the Charles and Louise Travers Department of Political Science at U.C. Berkeley; the Institute of International Studies at U.C. Berkeley; the Institute on Global Conflict and Cooperation at the University of California; the National Science Foundation; the Center for International Security and Cooperation at Stanford University; and the Project on Managing the Atom and the International Security Program at the Belfer Center for Science and International Affairs at Harvard University.

As I was finishing the manuscript, I was also able to benefit from comments at several research seminars across the country. I am grateful for helpful feedback from participants at seminars at the Olin Institute for Strategic Studies at Harvard University, the Program on International Security Policy at the University of Chicago, and the Center for International and Security Studies at the University of Maryland.

Parts of this book previously appeared in academic journal articles. I thank Cambridge University Press and Sage Publications for the permission to reprint this material.[2]

For helping this book through the publication process, I am grateful to Roger Haydon, the editors of the Cornell Studies in Security Affairs Series, and the staff at Cornell University Press. I also thank two anonymous reviewers for their constructive comments.

Last, but not least, I thank my family. My parents, Mark and Barb Kroenig, and my siblings, Brad and Julie Kroenig, have been an endless source of support, encouragement, and inspiration in everything that I have ever done. This book is dedicated to them.

---

2. Parts of this book previously appeared in Matthew Kroenig, "Exporting the Bomb: Why States Provide Sensitive Nuclear Assistance," *American Political Science Review*, Vol. 103, No. 1 (February 2009), pp. 113–133, reprinted with permission. For chapter 5, the final, definitive version of this paper has been published in Matthew Kroenig, "Importing the Bomb: Sensitive Nuclear Assistance and Nuclear Proliferation," *Journal of Conflict Resolution*, Vol. 52, No. 2 (April 2009), pp. 161–180, by SAGE Publications Ltd., All rights reserved. ©. The article is available at http://jcr.sagepub.com/cgi/content/abstract/53/2/161.

Abbreviations and Acronyms

AEC	Atomic Energy Commission (United States)
BOG	Board of Governors (International Atomic Energy Agency)
CEA	Commissariat à l'Énergie Atomique (France)
CFSD	Central Files State Department
CIA	Central Intelligence Agency (United States)
FLN	Front de Libération Nationale
HEU	highly enriched uranium
IAEA	International Atomic Energy Agency
IAEC	Israeli Atomic Energy Commission
IMF	International Monetary Fund
ISI	Inter-Services Intelligence (Pakistan)
JFKL	John F. Kennedy Library
KRL	Khan Research Laboratories
MID	Militarized Interstate Dispute
MW	megawatt
NARA	National Archives and Records Administration
NATO	North Atlantic Treaty Organization
NIE	National Intelligence Estimate
NPT	Nuclear Nonproliferation Treaty
NSC	National Security Council (United States)
NSG	Nuclear Suppliers Group
NTI	Nuclear Threat Initiative
Pinstech	Pakistan Institute of Science and Technology

PRC	People's Republic of China
PSI	Proliferation Security Initiative
ReLogit	Rare Events Logistic Regression
SGN	Saint-Gobain Techniques Nouvelles
SNIE	Special National Intelligence Estimate
TRA	Taiwan Relations Act
USNA	United States National Archives, College Park, Maryland
UN	United Nations
UNGA	United Nations General Assembly
URENCO	Uranium Enrichment Corporation
USSR	United Soviet Socialist Republics
WMD	weapons of mass destruction

EXPORTING THE BOMB

INTRODUCTION
The Problem of Nuclear Assistance

Nuclear proliferation poses a grave threat to international peace and security. For this reason, politicians, policymakers, and academics worry that nuclear-capable states will provide sensitive nuclear assistance to other states or terrorist networks, further fueling the spread of nuclear weapons. For example, following North Korea's nuclear test in October 2006, George W. Bush declared that "the transfer of nuclear weapons or material by North Korea to states or non-state entities would be considered a grave threat to the United States, and we would hold North Korea fully accountable of the consequences of such action."[1]

The empirical record provides reason for such concern. While states, to the best of our knowledge, have never provided sensitive nuclear assistance to terrorists, they have repeatedly transferred sensitive nuclear materials and technology to other states. From 1959 to 1965, France provided Israel with sensitive nuclear assistance; a mere two years after the end of French assistance, Israel is believed to have constructed its first nuclear weapon. This story, about one state helping another to develop nuclear weapons, is neither unique nor confined to the distant past. China assisted Pakistan with its nuclear program in the early 1980s with a package that included uranium enrichment technology, weapons-grade uranium, and a nuclear weapon design. Shortly thereafter, Pakistan is believed to have assembled its first nuclear bomb. More recently, from 1987 to 2002, Pakistan, with the help of Pakistani nuclear scientist A. Q. Khan, distributed sensitive

---

1. "President Bush's Statement on North Korea Nuclear Test," October 9, 2006.

nuclear technology and materials to Iran, Libya, and North Korea.[2] Since the end of this cooperation in 2002, Libya has agreed to give up its nuclear program, but North Korea has already tested its first nuclear device, and Iran is making steady progress on its own nuclear capability. These are a few of the many important cases of sensitive nuclear assistance that have contributed to the proliferation of nuclear weapons.

Yet, there is significant variation in the patterns of sensitive nuclear assistance. While some nuclear-capable states repeatedly provide sensitive nuclear assistance, many others refrain from transferring sensitive nuclear materials and technology altogether. Indeed, it is puzzling that states would transfer materials and technology that could help other states acquire nuclear weapons, the world's most destructive weapons, weapons that could one day threaten the suppliers' very existence.

This raises an interesting question about the motivations of the nuclear suppliers: Why do states provide sensitive nuclear assistance to nonnuclear weapon states, contributing to the international spread of nuclear weapons? This is the central question that this book will address. By "sensitive nuclear assistance," I mean the transfer of nuclear materials and technologies directly relevant to a nuclear weapons program. This includes nuclear weapon designs, weapons-grade fissile material, and sensitive nuclear fuel-cycle facilities. This definition distinguishes sensitive nuclear assistance from civilian nuclear cooperation related to basic nuclear science and technology or the production of nuclear energy.

Despite the obvious real-world importance of sensitive nuclear assistance, international relations scholars have not yet addressed this topic. Academic analysts have explained why states want nuclear weapons, why states transfer conventional arms, and the effect of nuclear proliferation on the probability of war and crisis behavior but have not carefully examined the supply side of nuclear proliferation.[3]

---

2. Pakistani assistance to Iran, Libya, and North Korea from 1987 to 2002 was state-sponsored by any reasonable definition of the term. As we will see in chapter 4, senior government officials, including heads of state and chiefs of army staff, actively supported the policy of nuclear transfer. See, for example, Gordon Corera, *Shopping for Bombs* (Oxford: Oxford University Press, 2006).

3. On state demand for nuclear weapons, see, for example, Scott D. Sagan, "Why Do States Build Nuclear Weapons?" *International Security*, Vol. 21, No. 3 (Winter 1996/97), pp. 54–86. On conventional arms transfers, see, for example, Andrew J. Pierre, *The Global Politics of Arms Sales* (Princeton, NJ: Princeton University Press, 1982). For a seminal work on the consequences of nuclear proliferation, see Scott D. Sagan and Kenneth N. Waltz., eds., *The Spread of Nuclear Weapons* (New York: W.W. Norton, 1995). For existing scholarly work on the supply of nuclear proliferation, see, for example, Kroenig, "Exporting the Bomb"; and Matthew Fuhrmann, "Taking a Walk on the Supply Side," *Journal of Conflict Resolution*, Vol. 53, No. 2 (April 2009). For examinations of how countries respond *after* another country has acquired nuclear weapons, see Steve E. Miller, "Assistance to Newly Proliferating Nations," in Robert D. Blackwill and Albert Carnesale, eds., *New Nuclear Nations*

Media reports and policy analyses of specific cases of sensitive nuclear assistance tend to assume that nuclear transfers are driven by economic considerations. Analysts claim that states in dire economic circumstances export sensitive nuclear materials and technology—and ignore the potential security risk—because they are in search of much-needed hard currency.

This book challenges this conventional wisdom. States that provide sensitive nuclear assistance understand the strategic consequences of their behavior quite well. In fact, state decisions to provide sensitive nuclear assistance are the result of a coherent, strategic logic.

The strategic theory of nuclear proliferation presented in this book is derived from a simple insight that is grounded in the nuclear deterrence literature: the spread of nuclear weapons threatens powerful states more than it threatens weak states.

Power-projecting states, states with the ability to project conventional military power over a particular target state, have a lot to lose when that target state acquires nuclear weapons. In interactions with a nonnuclear weapon state, power-projecting states can use their conventional military power to their advantage; they can threaten, or promise to protect, that particular state. Once that state acquires nuclear weapons, however, this strategic advantage is certainly placed at risk and may be fully lost. For these reasons, power-projecting states fear nuclear proliferation to both allied and enemy states. While the threat of nuclear proliferation is greatest when nuclear weapons are acquired by enemy states, nuclear proliferation, even to friendly states, can cause many problems for power-projecting states. Leaders in power-projecting states are concerned that nuclear proliferation will deter them from using military force to secure their interests, reduce the effectiveness of their coercive diplomacy, trigger regional instability that could engulf them in conventional conflict, weaken the integrity of their alliance structures, dissipate their strategic attention, and set off further nuclear proliferation within their spheres of influence.

On the other hand, non-power-projecting states, states that lack the ability to project power over a particular target state, do not incur these strategic costs when that target state acquires nuclear weapons. This is again true whether nuclear weapons spread to allied or enemy states. Non-power-projecting states are too weak to threaten hostile states and too weak to promise protection to friendly states, so they do not lose these strategic advantages as nuclear weapons spread. Because they lack the benefits afforded by a viable military option, the

---

(New York: Council on Foreign Relations Press, 1994), pp. 97–131; and Peter Douglas Feaver and Emerson M. S. Niou, "Managing Nuclear Proliferation: Condemn, Strike, or Assist," *International Studies Quarterly*, Vol. 40, No. 2 (Summer, 1996), pp. 209–233.

spread of nuclear weapons does not further undermine their strategic position. Their relative weakness precludes them from using military force to secure their interests, using military coercion as a tool of diplomacy, intervening in regional crises, extending security guarantees as a means to cement their alliance structures, needing to monitor new nuclear states, or needing to worry about further nuclear proliferation beyond their limited spheres of influence. Of course, non-power-projecting states may have other reasons to fear nuclear proliferation, but unlike for power-projecting states, the spread of nuclear weapons does not serve to constrain their conventional military freedom of action. For this reason, non-power-projecting states are on average less threatened by nuclear proliferation.

From this basic insight, this book derives three hypotheses about the conditions under which states will be more or less likely to provide sensitive nuclear assistance. First, because nuclear proliferation constrains states' conventional military freedom of action, the better able a state is to project power over a potential nuclear recipient, the less likely it is to provide sensitive nuclear assistance. States do not wish to constrain themselves. Second, precisely because nuclear proliferation constrains power-projecting states, states are more likely to provide sensitive nuclear assistance to states with which they share a common enemy. By providing sensitive nuclear assistance to these states, a nuclear supplier can impose particularly high strategic costs on powerful rivals. Finally, because superpowers, states with global, force-projection capabilities, are threatened by nuclear proliferation anywhere in the international system, they pressure other states in an attempt to dissuade sensitive nuclear transfers. States that are vulnerable to superpower pressure are less likely to provide sensitive nuclear assistance.

In short, states provide sensitive nuclear assistance for strategic reasons. They are more likely to export sensitive nuclear materials and technology when it would have the effect of constraining an enemy and less likely to do it when it would threaten themselves.

Alternative explanations cannot account for the patterns of sensitive nuclear transfers. Economic explanations may be intuitively appealing, but I will show that they are not met with empirical support. There is no discernable relationship between a country's economic circumstances and the likelihood that it will provide sensitive nuclear assistance. This is not to say that economic motivations are irrelevant to state decisions to transfer sensitive nuclear materials and technology. In some cases in which sensitive nuclear transfers occurred, the nuclear suppliers sought some form of payment. What the findings of this book make clear, however, is that states are unlikely to pursue economic gains when the result undermines their own security. States may still seek financial benefits when they export sensitive nuclear technology, but they are only likely to do so when such behavior is consistent with an underlying strategic logic.

Theories rooted in beliefs about the power of international institutions or related explanations about international norms cannot provide a satisfactory explanation for the causes of sensitive nuclear assistance. My finding is that the institutions of the nuclear nonproliferation regime have not exerted a consistent constraining effect on the nuclear export behavior of member states. Moreover, I find no support for the existence of an international norm against nuclear proliferation that discourages states from transferring sensitive nuclear material and technology. This is not to say that the nuclear nonproliferation regime has not served its purpose, however. While the regime may have failed to constrain sensitive nuclear transfers, I provide evidence that the NPT has played a role in retarding the international spread of nuclear weapons.

An explanation that centers on nuclear weapons possession cannot account for the patterns of sensitive nuclear assistance either. "Capable nuclear suppliers" are states that have the ability to provide sensitive nuclear assistance. Capable nuclear suppliers include states that possess nuclear weapons, like France, Pakistan, and the United States, as well as states, such as Brazil, Germany, and Japan, that have mastered parts of the nuclear-fuel cycle but have not developed nuclear weapons themselves. Some have argued that nuclear weapon states should be particularly opposed to nuclear proliferation because they have an incentive to prevent other countries from gaining entry into the nuclear club. Applied to the problem of sensitive nuclear assistance, we may expect that nuclear weapon states will be less likely to provide sensitive nuclear assistance than capable nuclear suppliers that lack nuclear weapons themselves. The evidence does not support this argument. Nuclear weapon states frequently help other states acquire nuclear weapons.

This book examines the consequences, as well as the causes, of sensitive nuclear assistance. When sensitive nuclear assistance occurs, what results? What is the relationship between sensitive nuclear assistance and the spread of nuclear weapons? I show that sensitive nuclear assistance is a powerful cause of nuclear proliferation. States that import sensitive nuclear materials and technology are more likely to acquire nuclear weapons than are similar states that do not receive outside help. Assistance from a more advanced nuclear state can help potential nuclear proliferators overcome the common technical and strategic hurdles they face as they attempt to develop nuclear weapons. The strong relationship between sensitive nuclear assistance and nuclear proliferation holds even after other factors thought to contribute to nuclear proliferation, such as living in a dangerous neighborhood with other nuclear-armed states or industrial capacity, are taken into account. This means that explaining sensitive nuclear assistance is a necessary first step if we hope to get a handle on how and why nuclear weapons spread. The key contribution of this book, therefore, is in providing empirical evidence for the first theoretical explanation of sensitive nuclear assistance.

I also address several other important issues in international relations theory and nuclear nonproliferation policy. First, I present a new, supply-side approach to understanding the causes of nuclear proliferation. The existing scholarship on nuclear proliferation has paid overwhelming attention to the demand side of nuclear proliferation. Explaining why states want nuclear weapons is important, but it is only part of the proliferation picture. After all, whether or not states want nuclear weapons is often immaterial, because other states take actions designed to assist or impede them. At one extreme, as we shall see, states are willing to provide sensitive nuclear assistance to help additional states acquire nuclear weapons. At the other extreme, states are willing to apply tough measures, including the use of military force, to stop the spread of nuclear weapons. If we are to understand the causes of nuclear proliferation, it is necessary to understand this supply side of nuclear proliferation. I examine the specific problem of sensitive nuclear transfers. Further research could examine other supply-side actions that states take to assist or impede nuclear proliferation in other states, such as voting on nuclear nonproliferation measures in international bodies, applying sanctions against nuclear proliferators, and supporting the use of military force against other states' nuclear programs.

Second, I provide an original understanding of the consequences of nuclear proliferation. The literature on nuclear proliferation effects has long been dominated by the debate between "proliferation optimists" and "proliferation pessimists."[4] Proliferation optimists argue that, when it comes to the spread of nuclear weapons, "more may be better" because nuclear weapons increase the cost of conflict, deterring leaders from engaging in war against nuclear-armed states.[5] The spread of nuclear weapons, in the optimists' conception, has a pacifying effect on international politics, leading to international stability. On the other hand, proliferation pessimists argue that "more will be worse" because more nuclear weapons in the hands of more states increases the chance of preventive wars, crisis instability, and accidental nuclear detonation.[6] According to the pessimists, nuclear proliferation contributes to greater levels of international instability.

The optimism/pessimism debate has done much to illuminate our understanding of the consequences of the spread of nuclear weapons. The existing scholarship, however, has been preoccupied with the study of the system-level

---

4. For the landmark work in this debate, see Sagan and Waltz, eds., *Spread of Nuclear Weapons*.

5. See, for example, Kenneth N. Waltz, "More May Be Better," in Sagan and Waltz, eds., *Spread of Nuclear Weapons*.

6. See, for example, Scott D. Sagan, "More Will Be Worse," in Sagan and Waltz, eds., *Spread of Nuclear Weapons*.

effects of nuclear proliferation. In particular, these scholars have examined whether nuclear proliferation increases or decreases the stability of the international system. For this reason, the existing scholarship has devoted less attention to unit-level effects. In other words, both optimists and pessimists focus on how nuclear proliferation affects the world, but not on how it affects the individual states that inhabit that world. Scholars have, of course, examined how nuclear proliferation will affect the nuclear proliferators themselves, but they have not systematically examined how nuclear proliferation in one state will affect other states.[7] Instead, the overarching concern of the optimism/pessimism debate has been to explain the relationship between nuclear proliferation and international stability. Due to this systemic focus, optimists and pessimists do not examine whether nuclear proliferation may affect different types of states differently. For instance, Waltz and Sagan tangle over whether the spread of nuclear weapons is good or bad, but they never seriously consider whether nuclear proliferation may be good for some states and bad for others.

In a significant departure from existing approaches, this book proposes a theory of nuclear proliferation that examines the unit-level, as well as the differential, effects of nuclear proliferation. It argues that nuclear proliferation threatens some states more than others and that the threat posed by nuclear proliferation depends on a state's ability to project military power. This book shows that power-projecting states are threatened by nuclear proliferation primarily because the spread of nuclear weapons constrains their conventional military freedom of action. On the other hand, states that lack a power-projection capability are less threatened by nuclear proliferation and, under certain circumstances, can even benefit from it. Pessimists argue that the spread of nuclear weapons is bad, optimists argue that it may be good—I put forward the argument that it depends: the spread of nuclear weapons is bad for power-projecting states and may be good for non-power-projecting states.

It is of course the case that many small powers have historically opposed nuclear proliferation, and weak states may have other reasons to fear the international spread of nuclear weapons. These facts do not directly challenge the central point. While powerful states' opposition to nuclear proliferation is determined by their strategic environment, the very different structural position of

---

7. The large literature on why states build nuclear weapons focuses at the unit level but asks a different question. This research does not examine how nuclear proliferation to some states will affect other states. Rather, it considers how states themselves will or will not benefit from pursuing or acquiring nuclear weapons and the implications of this on states' demand for nuclear weapons. See, for example, Erik Gartzke and Matthew Kroenig, "A Strategic Approach to Nuclear Proliferation," *Journal of Conflict Resolution*, Vol. 53, No. 2 (April 2009).

weak states provides them with greater freedom to oppose, or sometimes even support, the diffusion of nuclear weapons.

This novel approach promises to reinvigorate the study of the consequences of nuclear proliferation by establishing a research agenda on the differential effects of nuclear proliferation. Future research can identify the factors, other than power projection, that shape the degree to which states will be threatened by the spread of nuclear weapons.

Finally, the argument I propose has important implications for nuclear nonproliferation policy. In his 2009 annual report to Congress on the projected threats to the national security of the United States of America, Director of National Intelligence Dennis C. Blair assessed that the possibility of nuclear proliferation in various countries poses a great threat to U.S. national security.[8] This was not a new recognition. Nuclear proliferation has been considered one of the top threats to U.S. national security for decades. In response to the continuing threat of nuclear proliferation, the United States has implemented a number of policies designed to stop states from transferring sensitive nuclear materials and technology. The effective execution of these and other nonproliferation policies requires an accurate assessment of which states are most likely to export sensitive nuclear materials and technology. Correctly understanding the conditions under which states provide sensitive nuclear assistance matters not simply for academic analysts of nuclear proliferation, but also for those who wish to prevent nuclear proliferation in the future.

Plan of the Book

This book is divided into six chapters. The next chapter, chapter 1, presents the central argument of the book. It examines the differential effects of nuclear proliferation and spells out their implications for the problem of sensitive nuclear assistance. The chapter explains how the spread of nuclear weapons disproportionately threatens power-projecting states, leading to three strategic conditions under which states are most likely to export sensitive nuclear materials and technologies. The chapter also develops competing explanations of sensitive nuclear assistance that will be tested against the power-based theory.

Chapters 2 through 5 constitute the empirical core of the book. Chapter 2 presents statistical results that reveal the correlates of sensitive nuclear assistance. The statistical analysis draws on an original sensitive nuclear assistance dataset.

---

8. Dennis C. Blair, "Annual Threat Assessment of the Intelligence Community for the Senate Select Committee on Intelligence," February 12, 2009, unclassified Statement for the Record.

The dataset contains yearly information for all capable nuclear suppliers and potential nuclear recipients in the international system from 1951 to 2000.

In chapters 3 and 4, I use pattern-matching and process-tracing methods to examine the causal logic of the argument. Chapter 3 compares French assistance to Israel's nuclear weapons program from 1959 to 1965 with U.S. efforts to thwart Israel's nuclear development in the same time period. Chapter 4 tests the limits of the argument through an examination of several brief case studies. This chapter considers additional cases of sensitive nuclear assistance, as well as cases in which sensitive nuclear assistance could have occurred, but did not.

Chapter 5 assesses the consequences of sensitive nuclear assistance. Does sensitive nuclear assistance lead to nuclear proliferation? Drawing on case studies and quantitative analysis, this chapter demonstrates that states that receive sensitive nuclear assistance are more likely to acquire nuclear weapons than are similar states that do not receive outside help.

The conclusion provides a review of the argument and turns to the implications of the findings for theory and practice. The main thrust of the conclusion focuses on the implications of the argument for international relations theory. This analysis advances important scholarly debates on both the causes and the consequences of nuclear proliferation. The conclusion finishes with a discussion of the applications of the argument to U.S. nonproliferation policy. It considers the states that are at risk of becoming the next nuclear suppliers and explores whether they will provide sensitive nuclear assistance to rogue states or terrorist networks.

1
EXPLAINING NUCLEAR ASSISTANCE

This chapter develops a strategic theory of nuclear proliferation and applies it to the problem of sensitive nuclear assistance. In short, I will argue that nuclear proliferation threatens some states more than others and that the threat posed by nuclear proliferation depends on a state's ability to project military power. States that have the ability to project military power over a particular target state, states that I call "power-projecting states," are most threatened by nuclear proliferation to that particular state because nuclear proliferation in that instance will constrain their conventional military freedom of action. On the other hand, states that lack the ability to project power over a particular target state, states that I call "non-power-projecting states," are less threatened by nuclear proliferation to that state. This simple strategic logic gives rise to three conditions that shape the probability of sensitive nuclear assistance.

The Elements of the Argument

Before detailing my theoretical argument, I will define the core concepts that will be used throughout the study. First, I explain the dependent variable: sensitive nuclear assistance. Then, I develop the key explanatory variable: power-projecting state.

Sensitive Nuclear Assistance

Sensitive nuclear assistance is the state-sponsored transfer of the key materials and technologies necessary for the construction of a nuclear weapons arsenal to a nonnuclear weapon state. Sensitive nuclear assistance takes three forms. States provide sensitive nuclear assistance when they assist nonnuclear weapon states in the design and construction of nuclear weapons; transfer significant quantities of weapons-grade fissile material to nonnuclear weapon states; or assist nonnuclear weapon states in the construction of facilities to produce weapons-grade fissile material.

These three forms of sensitive nuclear assistance require further elaboration. First, sensitive nuclear assistance may take the form of assistance in the design and construction of a nuclear weapon itself, such as the international transfer of a nuclear bomb design. Helping another state construct a functioning nuclear weapon is the most direct way to contribute to another state's nuclear weapons capability and should obviously be included in this definition of sensitive nuclear assistance.

Second, sensitive nuclear assistance may take the form of the interstate transfer of significant quantities of weapons-grade fissile material. Nuclear weapons experts have long recognized that a necessary, and also the most difficult, step in building a nuclear weapon is the acquisition of the weapons-grade fissile material that forms the core of the nuclear device.[1] International Atomic Energy Agency (IAEA) regulations assume that 8 kilograms of plutonium and 25 kilograms of weapons-grade highly enriched uranium (HEU) are sufficient for the construction of a nuclear device. States that transfer significant quantities of weapons-grade fissile material to a nonnuclear weapon state are providing sensitive nuclear assistance.

Third, sensitive nuclear assistance may also take the form of assistance in the construction of facilities to produce weapons-grade fissile material. Uranium enrichment and plutonium reprocessing are the two means by which states can produce weapons-grade fissile material, and nuclear weapons experts have long recognized that these fuel-cycle capabilities present a clear nuclear proliferation threat. For example, IAEA Director General Mohamed El-Baradei has averred, "Should a state with a fully-developed fuel-cycle capability decide, for whatever reason, to break away from its non-proliferation commitments, most experts believe it could produce a nuclear weapon within a matter of months."[2] Assistance on sensitive fuel-cycle facilities includes the construction of a complete uranium enrichment or plutonium reprocessing plant. Assistance related to sensitive fuel-cycle facilities also includes the transfer of key component parts

---

1. For a primer on nuclear weapons and their construction, see Rodney W. Jones and Mark G. McDonough with Toby F. Dalton and Gregory D. Koblentz, *Tracking Nuclear Proliferation* (Washington, D.C.: Carnegie Endowment for International Peace, 1998), pp. 317–322.

2. Mohamed El-Baradei, "Towards a Safer World," *Economist*, October 16, 2003.

for the construction of such facilities, such as centrifuges for uranium enrichment plants or hot cells for plutonium reprocessing plants. Assistance on uranium enrichment includes assistance on any of the various types of uranium enrichment processes, including: jet-nozzle, gaseous diffusion, gas-centrifuge, and laser-isotope enrichment.

Sensitive nuclear assistance includes the above forms of nuclear transfer whether or not they are provided under international safeguards. International safeguards allow for the international monitoring of nuclear facilities to detect the diversion of fissile materials. Safeguards are a notoriously thin reed on which to rest one's security, however. States have used the technological expertise gained at safeguarded facilities to develop parallel, unsafeguarded nuclear programs; circumvented safeguards provisions; expelled international inspectors; and withdrawn from international safeguards. For these reasons, transfers of sensitive nuclear materials and technology provided under international safeguards represent a deliberate, and potentially dangerous, state effort to aid substantially another state's nuclear program and are, therefore, included in this definition of sensitive nuclear assistance.

Sensitive nuclear assistance excludes other types of nuclear cooperation less relevant to the development of a nuclear weapons program. The provision of nonsensitive, civilian nuclear assistance, including scientific exchanges, assistance in the surveying and mining of natural uranium, the provision of reactor fuel and services, and the construction of research and power reactors, does not qualify as sensitive nuclear assistance. The line between sensitive and nonsensitive nuclear assistance is sometimes fuzzy, yet there is a widespread scientific consensus that sensitive fuel-cycle facilities, such as uranium enrichment facilities, represent a direct nuclear proliferation threat, while other less sensitive, civilian technologies are relatively resistant to proliferation. By drawing the line between sensitive and nonsensitive nuclear assistance at sensitive fuel-cycle facilities, my definition follows this preexisting consensus.

Sensitive nuclear assistance also excludes the transfer of dual-use industrial parts that could be used in the construction of sensitive fuel-cycle facilities but are not themselves sensitive nuclear materials or technologies. For example, in the 1980s and 1990s, German firms exported materials that Iraq used in the construction of its nuclear facilities.[3] These materials included electrical components, industrial pipelines, soft iron, and furnace equipment. For obvious

---

3. Jennifer Hunt Morstein and Wayne D. Perry, "Commercial Nuclear Trading Networks as Indicators of Nuclear Weapons Intentions," *Nonproliferation Review,* Vol. 7, No. 3 (Fall/Winter 2000), pp. 75–91; and Nuclear Threat Initiative, *Country Profiles.*

reasons, I do not define iron exports and other transfers of dual-use industrial materials as sensitive nuclear assistance.

Why states decide to help other states develop the platforms that could be used to deliver nuclear weapons, such as bombers, ballistic missiles, and submarines is an interesting question, but it is beyond the scope of this study.

Sensitive nuclear assistance also excludes transfers of sensitive nuclear technology to established nuclear weapon states or transfers that do not materially advance a state's nuclear program.[4] For example, the United States assisted Great Britain and France with nuclear weapons and strategic delivery vehicles during the Cold War, but the assistance between these countries occurred after they had already acquired nuclear weapons.[5] These cases, and others like them, are not considered instances of sensitive nuclear assistance because the empirical and theoretical puzzle motivating this research centers on why states help nonnuclear weapon states acquire nuclear weapons. Why nuclear weapon states trade nuclear technology among themselves is an interesting question but, again, beyond the scope of this study. Similarly, assistance to a nonnuclear weapon state related to a sensitive nuclear technology that the nonnuclear weapon state has already mastered does not qualify as sensitive nuclear assistance because it does not advance that state's ability to produce nuclear weapons. For example, Japan began operating domestic plutonium reprocessing facilities in 1977. Current French-Japanese cooperation on the construction of a new plutonium reprocessing facility in Japan is not counted as sensitive nuclear assistance because this assistance does not advance Japan's technical ability to produce nuclear weapons. French-Japanese cooperation related to nuclear weapon design, or other sensitive nuclear areas that Japan has not already mastered, however, would qualify as sensitive nuclear assistance.

The international smuggling of sensitive nuclear technology by substate actors such as individuals, firms, or bureaucracies without the government's knowledge or approval is not sensitive nuclear assistance as I define it. This state-centric focus allows for the study of the vast majority of sensitive nuclear transactions since the substate smuggling of sensitive nuclear materials and technology is empirically rare. There has been one case, or possibly two cases, of sensitive

---

4. For an analysis of the potential advantages and disadvantages of providing assistance designed to improve the safety and security of nuclear arsenals in new nuclear weapon states, see Miller, "Assistance to Newly Proliferating Nations," pp. 97–131.

5. Contrary to the belief of many, the United States did not provide meaningful sensitive nuclear assistance to Great Britain during the Manhattan Project. Rather, Washington attempted to keep London from the bomb and strove to preserve its nuclear monopoly. For more on U.S. nuclear cooperation with the United Kingdom and France and on why these cases are not considered instances of sensitive nuclear assistance, see appendices D and E.

nuclear assistance without direct state involvement.[6] This empirical finding supports the intuition that it is prima facie implausible that nuclear-capable states would not exert control over their most sensitive military technologies and that substate actors could effectively conduct large-scale, sensitive nuclear transfers without the knowledge and approval of senior government officials.[7]

This measurement of sensitive nuclear assistance does not gauge a state's nuclear nonproliferation policies more broadly. As was mentioned in the introduction, a state's stance on nuclear proliferation can be thought of as existing along a continuum from outright opposition to open support. At one extreme, states use preventive military strikes to halt nuclear proliferation, and at the other end of the spectrum, states provide generous nuclear assistance to encourage the further spread of nuclear weapons. In between are a range of options, including economic sanctions, diplomatic pressure, turning a blind eye, pro forma protests, and tacit support. Examining this entire spectrum of nuclear proliferation behavior could make for an interesting study, but is beyond the scope of the present book. Rather, I focus in on what is probably the most puzzling, and arguably the most important, proliferation behavior: the provision of sensitive nuclear assistance.

In sum, sensitive nuclear assistance includes the international transfer of nuclear weapon designs, weapons-grade fissile material, and sensitive fuel-cycle facilities to nonnuclear weapon states, and excludes everything else.

Power-Projecting States

Power-projecting states are states that have the ability to fight a full-scale, conventional, military, ground war on the territory of a potential target state. Non-power-projecting states are states that lack this capability. To qualify as a power-projecting state, a state does not need the capability to defeat the target state, but it must have a reasonable chance of seriously weakening the target state, even if the target state ultimately wins the war.[8]

---

6. An analysis of these cases would be interesting but is beyond the scope of the present work. A description of these cases is available in appendices D and E.

7. There have been a number of instances in which individuals have attempted to smuggle small amounts of radioactive materials across international borders, but none of these cases involved significant quantities of weapons-grade fissile material. For a list of these cases of nuclear smuggling, see Government Accountability Office, *Nuclear Nonproliferation*, May 2002.

8. This definition draws heavily from John Mearsheimer's definition of a great power. According to Mearsheimer, a great power is a state that possesses "sufficient military assets to put up a serious fight in an all-out conventional war against the most powerful state in the world. The candidate need not have the capability to defeat the leading state, but it must have some reasonable prospect of turning the conflict into a war of attrition that leaves the dominant state seriously weakened, even if that

The ability to bomb a state alone, without a corresponding ability to put boots on the ground in that state's territory, is not a sufficient power-projection capability as I define it. While air power can, in certain circumstances, provide states with coercive leverage, I start from the presumption, consistent with a strong intellectual tradition in international relations, security studies, and military analysis, that occupying and controlling territory is the key to power in international politics. As John Mearsheimer writes, "Armies are the central ingredient of military power, because they are the principal instrument for conquering and controlling territory—the paramount political objective in a world of territorial states."[9] Furthermore, there are compelling theoretical and empirical reasons to believe that air power alone, without an associated threat of ground invasion, does not provide a significant coercive advantage. For example, in a seminal study, Robert Pape demonstrates that strategic bombing rarely helps states achieve their foreign policy objectives.[10] Moreover, empirically and pragmatically, a definition of power projection that includes air strikes alone does not help to distinguish states' nuclear proliferation behavior. As we will see later in this book, there are states like France and China that may have enjoyed the ability to bomb every corner of the globe but were still perfectly willing to forfeit this advantage and provide sensitive nuclear assistance to other states time and time again. On the other hand, as we will see, states that possessed the ability to launch a ground invasion of another state were more careful to guard this strategic prerogative and were more likely to refrain from providing sensitive nuclear assistance. I therefore follow other theorists of international relations in emphasizing the importance of the ability to launch a ground invasion of another state in assessing a state's ability to project power over a particular target state.

Power projection can only be assessed through careful military analysis. What matters is whether the state in question possesses the geographical position, force posture, and military capabilities that would allow it to invade a particular state at a particular time in the face of a hostile defense. Such careful military analysis is performed in the case studies in chapters 3 and 4. Power-projection capability cannot be definitively calculated by simply aggregating and comparing standard measures of power, such as GDP, military spending, or population size. These measures do, however, provide a rough estimate of relative power between states and will be employed in the statistical analyses in chapter 2.

---

dominant state ultimately wins the war." John J. Mearsheimer, *The Tragedy of Great Power Politics* (New York: W.W. Norton, 2001), p. 5.
  9. Ibid., p. 43.
  10. Robert Pape, *Bombing to Win* (Ithaca, NY: Cornell University Press, 1995).

It is important to note that this definition of power is a relative one. States may be able to project power against some states, but not others. For example, China could fight a ground war in Vietnam as it proved in the Sino-Vietnamese War of 1979, but given geographical constraints, its force posture, and military capabilities, China could not plausibly launch a military invasion of Iran.[11] According to this definition, China is a power-projecting state in relation to Vietnam, but a non-power-projecting state vis-à-vis Iran. In each case, it is important to remember that power-projection capability is assessed in relation to a particular target state.

A Strategic Theory of Nuclear Proliferation

This section develops a strategic theory of nuclear proliferation. I begin by developing the argument that nuclear proliferation threatens power-projecting states more than it threatens non-power-projecting states. I then apply this insight to the problem of sensitive nuclear assistance to derive hypotheses about the conditions under which states will be more or less likely to provide sensitive nuclear assistance.

The Differential Effects of the Nuclear Revolution

The impact of nuclear proliferation, far from being uniform across states, depends on a state's position in the international system. Power-projecting states have the most to lose from nuclear proliferation because nuclear proliferation constrains these states in a number of ways. On the other hand, non-power-projecting states do not incur the same costs as nuclear weapons spread.

WHY MORE IS EVEN WORSE FOR POWER-PROJECTING STATES

The spread of nuclear weapons threatens power-projecting states primarily because it constrains their conventional military power. These states understand that the spread of nuclear weapons to states against which they have the option to use military force will erode a source of strategic advantage. While the threat is most severe when enemies acquire nuclear weapons, nuclear proliferation to friendly states still causes many problems for power-projecting states. Indeed,

---

11. It could be argued that that China's invasion of Vietnam was largely unsuccessful and demonstrated the limits of Chinese power. This may be true, but China still possessed the ability to project power over Vietnam as defined here. This definition of power-projection capability centers on whether a state can fight a ground war against a particular state but says little about the outcome of the conflict.

some of the strategic costs of nuclear proliferation detailed below are most salient when nuclear arms spread to close allies. Leaders in power-projecting states fear that nuclear proliferation might deter them from using military intervention to pursue their interests, reduce the effectiveness of their coercive diplomacy, trigger regional instability, undermine their alliance structures, dissipate their strategic attention, and set off further nuclear proliferation within their sphere of influence. These strategic costs are not as catastrophic as nuclear war, but they are costs that power-projecting states can count on incurring with near-certainty as nuclear weapons spread. Power-projecting states also consider other high-impact, low-probability consequences of nuclear proliferation, such as nuclear war, accidental nuclear detonation, or, in recent years, nuclear terrorism. But, evidence from their own internal, strategic assessments reveals that statesmen in power-projecting states are threatened by nuclear proliferation because they understand that it will constrain their conventional military freedom of action.

To develop the argument, I will draw on illustrative evidence from the strategic calculations of policymakers in power-projecting states. Much of this evidence will come from the U.S. experience with nuclear proliferation for two reasons. First, the United States enjoys global force-projection capabilities and is a power-projecting state, as I define it, in relation to every other state in the international system.[12] Second, abundant access to declassified and other archival materials provides excellent insight into how U.S. officials assess the threat posed by nuclear proliferation. To demonstrate that the constraining effects of nuclear proliferation extend beyond the United States, this section will also present available evidence from other power-projecting states. The Soviet Union during the Cold War possessed global force-projection capabilities and was a power-projecting state vis-à-vis every other state in the world. I will also provide evidence from power-projecting states that lack global force-projection capabilities. These examples will draw from a number of states that border potential target states. These states can project power against these target states by launching a ground invasion across a common border. Dyads of power-projecting states and potential target states considered here include: Egypt and Israel, India and Pakistan, Turkey and Iran, and South Korea and North Korea.

Deters Military Intervention. Policymakers in power-projecting states fear that nuclear weapons could deter them from using conventional military force to pursue their interests. This belief is consistent with much of the nuclear deterrence

---

12. On U.S. force-projection capabilities, see, for example, Barry R. Posen, "Command of the Commons: The Military Foundation of U.S. Hegemony," *International Security*, Vol. 28, No. 1 (Summer 2003), pp. 5–46.

literature that claims that nuclear weapons deter foreign invasion.[13] Power-projecting states can use force in an attempt to reduce the military capabilities, change the policies, or even overthrow the governments of threatening nonnuclear weapon states. When facing a nuclear power, however, direct military intervention becomes a much less attractive option. Power-projecting states are deterred from using their conventional military power against threatening nuclear weapon states, constraining their military freedom of action. Indeed, the benefit of nuclear deterrence is often thought to be the primary reason why states want nuclear weapons.[14]

Of course, nuclear deterrence may not always work. Nuclear-armed states, like Israel, have been attacked, and theorists of the stability/instability paradox claim that strategic nuclear deterrence could help to encourage low-level conflicts.[15] Still, nuclear weapons are widely regarded by policymakers and academics as having powerful deterrent effects. Even theorists of the stability/instability paradox admit that nuclear weapons impose constraints on the use of conventional military power because, while nuclear weapons may induce low-level conflict, states could still be deterred from engaging in high-level conflicts that could escalate to the nuclear level.[16]

The deterrent effects of nuclear proliferation on a state's conventional military power have long been recognized and feared by the leaders of power-projecting states. For example, a 1961 U.S. Joint Chiefs of Staff report concluded, "A nuclear China would only weaken Washington's influence in the region and its capabilities to intervene on behalf of its allies there."[17] Similarly, a 1963 U.S. National Intelligence Estimate (NIE) assessed that if China acquired nuclear weapons, "the U.S. would be more reluctant to intervene on the Asian mainland."[18] This view was shared by President John F. Kennedy, who "feared that even a minimal Chinese nuclear force could prevent U.S. military intervention" in China. Kennedy

---

13. On nuclear deterrence theory, see, for example, Thomas Schelling, *Arms and Influence* (New Haven: Yale University Press, 1966); Robert Jervis, *The Meaning of the Nuclear Revolution* (Ithaca, NY: Cornell University Press, 1989); and Robert Powell, *Nuclear Deterrence Theory* (Cambridge, MA: Cambridge University Press, 1990).

14. On why states want nuclear weapons, see, for example, Sagan, "Why Do States Build Nuclear Weapons?"

15. For the original formulation of the stability-instability paradox, see Glenn H. Snyder, "The Balance of Power and the Balance of Terror," in Paul Seabury, ed., *The Balance of Power* (San Francisco: Chandler, 1965). For a critique and an extension of the stability-instability paradox, see S. Paul Kapur, "India and Pakistan's Unstable Peace," *International Security*, Vol. 30, No. 2 (Fall 2005), pp. 127–152.

16. Ibid.

17. William Burr and Jeffrey T. Richelson, "Whether to Strangle the Baby in the Cradle," *International Security*, Vol. 25, No. 3 (Winter 2000), p. 61.

18. Director of Central Intelligence, Special National Intelligence Estimate (SNIE) 13-2-63, "Communist China's Advanced Weapons Program," July 24, 1963, as cited in Burr and Richelson, "Whether to Strangle," p. 66.

further noted that just a few missiles in Cuba "had a deterrent effect on us."[19] Partly for this reason, Kennedy thought that China's imminent nuclear ascendance to the nuclear club was "likely to be historically the most significant and worst event of the 1960s."[20]

U.S.-based analysts have continued to fear the effect of nuclear deterrence on U.S. conventional military might since Kennedy's time. A 1986 top-secret Central Intelligence Agency (CIA) assessment, *North Korea: Potential for Nuclear Weapons Development*, stated that a nuclear North Korea "would have the effect of deterring a U.S. response to a North Korean attack."[21] Indeed, many analysts suspect that one of the reasons that the United States invaded Iraq in the spring of 2003 and not North Korea, another state designated by President Bush as a member of the "axis of evil," was that the United States was deterred by North Korea's nuclear arsenal.[22] As the United States considers the very real possibility that Iran may soon acquire nuclear weapons, U.S. military planners are undoubtedly concluding that one of the primary strategic consequences of a nuclear-armed Iran will be that the United States will be deterred from using military force in the Middle East.

Other power-projecting states also recognize that nuclear weapons deter military intervention. Egyptian officials were adamantly opposed to nuclear proliferation in neighboring Israel in the 1960s because they believed it would constrain Egypt's conventional military freedom of action. Avner Cohen explains that, according to Egyptian military assessments, "A soon-to-be-built Israeli nuclear weapon would put the Egyptian military in an inferior position, negating Egypt's conventional superiority."[23] At present, strategic thinkers in Turkey oppose nuclear proliferation in neighboring Iran because they believe that an Iranian bomb could threaten the conventional military balance between Turkey and Iran. Mustafa Kibaroglu writes that today, a rough "parity exists between [Iran and Turkey] in geographical location, demographic structure, and military capability," adding that "should Iran develop nuclear weapons capability, the balance may tip dramatically in favor of Iran."[24]

---

19. Marc Trachtenberg, *A Constructed Peace* (Princeton, NJ: Princeton University Press, 1999), p. 320.

20. James Fetzer, "Clinging to Containment," in Thomas G. Paterson, ed., *Kennedy's Quest for Victory* (New York: Oxford University Press, 1989), p. 182.

21. Central Intelligence Agency, *North Korea: Potential for Nuclear Weapons Development*, September 1986, p. vi, as quoted in Richelson, *Spying on the Bomb* (New York: W.W. Norton, 2006), p. 347.

22. See, for example, Fred Kaplan, "The Unspeakable Truth," *Slate*, January 7, 2003.

23. Avner Cohen, *Israel and the Bomb* (New York: Columbia University Press, 1998), p. 265.

24. Mustafa Kibaroglu, "Good for the Shah, Banned for the Mullahs," *Middle East Journal*, Vol. 60, No. 2 (Spring 2006), pp. 207–235.

Similarly, Indian officials opposed nuclear proliferation in Pakistan because they feared that a Pakistani nuclear arsenal would constrain India's more powerful military.[25] Indian security strategy in relation to Pakistan had long rested on a conventional military superiority that enabled India to threaten the territorial integrity of Pakistan without the fear of a credible retaliatory threat. But in the 1980s, Indian officials, including General K. Sundarji, chief of the army staff, feared Pakistan's nuclear program primarily because they believed that a nuclear arsenal in Pakistan would deter an Indian conventional attack, undermining India's military advantage.[26] Analysts have concluded that Sundarji was justified in this fear and that Pakistan's nuclear arsenal has had precisely this effect. Ashley Tellis writes that the primary effect of Pakistan's nuclear arsenal has been to "significantly circumscribe India's political and military freedom of action.... In effect, Pakistan—the traditionally weaker adversary—has now neutralized India's conventional and strategic advantages."[27]

Reduces Effectiveness of Coercive Diplomacy. For power-projecting states, nuclear proliferation reduces the effectiveness of coercive diplomacy. Nuclear proliferation not only deters power-projecting states from using military force against adversaries, it undermines the credibility of their threats to use military force. Students of coercive diplomacy maintain that the effectiveness of deterrence and compellence policies hinges on the credibility of their associated threats.[28] Adversaries are unlikely to be influenced by a threat that they believe will never be carried out. As the spread of nuclear weapons makes it difficult for power-projecting states to use military force, it also reduces their adversaries' estimations of the probability that they will follow through on threats to use force. The presence of nuclear weapons places a limit on how hard leaders in power-projecting states believe they can push in a crisis and, accordingly, power-projecting states limit their aims and means in conflicts with nuclear-armed adversaries. Power-projecting states may be forced to consider the redeployment of military forces and bases beyond the range of the new nuclear weapon state's arsenal to minimize their military vulnerability in potential crises. Power-projecting states may also be more likely to capitulate in political conflicts of interest against nuclear-armed

---

25. See Lieutenant General K. Sundarji, ed., "Effects of Nuclear Asymmetry on Conventional Deterrence," *Combat Papers* (Mhow), No. 1 (April 1981); Lieutenant General K. Sundarji, ed., "Nuclear Weapons in the Third World Context," *Combat Papers* (Mhow), No. 2 (August 1981); George Perkovich, *India's Nuclear Bomb* (Berkeley: University of California Press, 1999), p. 230; and Ashley J. Tellis, *India's Emerging Nuclear Posture* (Santa Monica, CA: RAND, 2001).
26. Ibid.
27. Tellis, *India's Emerging Nuclear Posture*, pp. 45–46.
28. On credibility and coercion, see, for example, Powell, *Nuclear Deterrence Theory;* and Schelling, *Arms and Influence.*

powers. Indeed, recent quantitative analyses have demonstrated that states are less likely to prevail in international disputes against nuclear-armed states.[29] As a power-projecting state backs down in confrontations with a new nuclear-armed state, the influence of the new nuclear weapon state in the geographical region is enhanced at the expense of the power-projecting state. At the extreme, policy-makers in power-projecting states worry that nuclear proliferation will allow the new nuclear weapon state to dominate their geographical region. The fear that nuclear weapons alone will allow a state to dominate a geographical region may be exaggerated, but nuclear weapons do appear to shift the bargaining space in favor, and increase the strategic influence, of their possessor.

Statesmen in power-projecting states recognize that nuclear proliferation could lead to a reduction in their bargaining power and regional influence. A 1963 U.S. NIE noted that a nuclear-armed China "would feel very much stronger and this mood would doubtless be reflected in their approach to conflicts...the tone of Chinese policy would probably become more assertive."[30] In their newfound assertiveness, U.S. analysts feared that a nuclear-armed China would be less willing to concede to U.S. demands and was sure "to exploit nuclear weapons for this end."[31] President Kennedy was convinced that China was "bound to get nuclear weapons, in time, and that from that moment on they will dominate South East Asia."[32] Considering the effect of nuclear proliferation more broadly, the Gilpatric Committee, a special committee set up by President Lyndon B. Johnson to analyze the implications of nuclear proliferation for U.S. foreign policy, assessed that nuclear proliferation could "eventually lead to the withdrawal of U.S. and Soviet forces from regions populated with new nuclear powers." The nuclear arming of China would lead to a reduction of U.S. influence in East Asia, which could then "fall under Chicom [Communist China] hegemony."[33] While some of the worst-case scenarios envisioned by U.S. officials did not come to pass, they were correct to believe that the United States would seek to avoid militarized disputes against a nuclear-armed China. Scholars have

---

29. Gartzke and Kroenig, "A Strategic Approach to Nuclear Proliferation"; Kyle Beardsley and Victor Asal, "Winning with the Bomb," *Journal of Conflict Resolution*, Vol. 53 No. 2 (April 2009); and Erik Gartzke and Dong-Joon Jo, "Bargaining Nuclear Proliferation and Interstate Disputes," *Journal of Conflict Resolution*, Vol. 53, No. 2 (April 2009).

30. Director of Central Intelligence (DCI), NIE 13-2-63, "Communist China's Advanced Weapons Program," p. 10, quoted in Burr and Richelson, "Whether to Strangle," p. 66.

31. "China as a Nuclear Power (Some Thoughts prior to the Chinese Test)," author unknown, October 7, 1964, National Security Files (hereafter cited as NSF), Committee on Non-Proliferation, box 5, p. 2, Lyndon Baines Johnson Library, Austin, Texas (hereafter cited as LBJL), quoted in Francis J. Gavin, "Blasts from the Past," *International Security*, Vol. 29, No. 3 (Winter 2004/2005), p. 104.

32. Trachtenberg, *Constructed Peace*, p. 320.

33. "Probable Consequences: Permissive or Selective Proliferation," author and date unknown, PPRG, box 11, John F. Kennedy Library (JFKL), quoted in Gavin, "Blasts from the Past," p. 110.

noted, for example, that the United States became much less willing to challenge China's core security interests after Beijing acquired the bomb.[34]

Similarly, in recent years, U.S. officials and U.S.-based analysts have concluded that nuclear proliferation would lead to constraints on U.S. influence and allow hostile states to gain greater sway in vital strategic regions. Barry Posen has argued that if Saddam Hussein had possessed nuclear weapons during the First Gulf War, the United States still could have gone to war against Iraq, but the United States would have been forced to limit its war aims and means.[35] The administration of President George W. Bush also feared that nuclear proliferation in Iraq could lead to a shift in bargaining power. In the runup to the Second Gulf War, President Bush warned that if Saddam Hussein acquired nuclear weapons, he "would be in a position to dominate the Middle East."[36] Today, U.S. officials maintain that a nuclear-armed Iran would reduce U.S. leverage, giving Iran greater influence over Middle Eastern politics. Peter Brookes, a U.S. deputy assistant secretary of defense in George W. Bush's administration, predicted that a nuclear-armed Tehran would become "the predominant state in the Middle East, replacing the U.S. as the region's power broker and lording over its Sunni Arab neighbors."[37]

Other power-projecting states also find that nuclear proliferation will reduce their diplomatic advantages and increase the influence of new nuclear weapon states. The Soviet Union feared that nuclear proliferation in Israel would reduce Moscow's strategic influence in the Middle East.[38] Egypt was adamantly opposed to nuclear proliferation in neighboring Israel in the 1960s because, according to Avner Cohen, Egyptian officials believed that an Israeli bomb would have the effect of "reducing the influence of the Egyptian armed forces."[39] At present, Turkey opposes nuclear proliferation in neighboring Iran because they believe that an Iranian bomb would enhance Tehran's coercive bargaining power and regional influence.[40]

Triggers Regional Instability. Nuclear proliferation can embolden new nuclear states, triggering regional instability that could potentially threaten the interests

---

34. Ta Jen Liu, *U.S.-China Relations, 1784–1992* (New York: University Press of America, 2002); and Appu K. Soman, *Double-Edged Sword* (New York: Praeger Publishers, 2000).
35. Barry Posen, "U.S. Security Policy in a Nuclear-Armed World (Or: What If Iraq Had Had Nuclear Weapons?)," *Security Studies*, Vol. 6, No. 3 (1997), pp. 1–31.
36. "President Bush Outlines Iraqi Threat," October 7, 2002.
37. Peter Brookes, "Iran Emboldened," *Armed Forces Journal*, April 2, 2007.
38. Isabella Ginor and Gideon Remez, *Foxbats over Dimona* (New Haven: Yale University Press, 2007).
39. Cohen, *Israel and the Bomb*, p. 265.
40. Kibaroglu, "Good for the Shah."

of power-projecting states and even ensnare them in regional disputes. Recent scholarly analyses have demonstrated that, after controlling for other relevant factors, nuclear weapon states are more likely to engage in conflict than non-nuclear weapon states and that this aggressiveness is more pronounced in new nuclear states that have less experience with nuclear diplomacy.[41] Similarly, research on internal decision-making in Pakistan reveals that Pakistani policymakers may have been emboldened by the acquisition of nuclear weapons, which encouraged them to initiate militarized disputes against India.[42]

The threat that regional instability poses to power-projecting states is different from the concern about international instability expressed by the proliferation pessimists. Pessimists assume that international instability is bad in and of itself—and they may be right. But power-projecting states have a different concern. They worry that nuclear proliferation will set off regional instability and that, because they have the ability to project power over the new nuclear weapon state, they will be compelled to intervene in a costly conflict. Power-projecting states could feel compelled to act as a mediator between nuclear-armed disputants, provide conventional military assistance to one of the parties in the dispute, or even be drawn into the fighting themselves because they have the ability to put boots on the ground in the new nuclear state.

The possibility that nuclear proliferation could spur conflict entangling powerful states is a negative consequence of nuclear proliferation that could occur whether nuclear weapons spread to friends or foes. Indeed, nuclear weapons in the hands of allies could trigger regional conflict that requires intervention just as easily as nuclear arms in the possession of rival states.

For these reasons, power-projecting states worry about the effect of nuclear proliferation on regional stability. As we will see in chapter 3, U.S. officials feared that nuclear proliferation in Israel might embolden Israel against its Arab enemies or entice Arab states to launch a preventive military strike on Israel's nuclear arsenal. In a 1963 NIE on Israel's nascent nuclear program, the consensus view of the U.S. intelligence community was that if Israel acquired nuclear weapons, "Israel's policy toward its neighbors would become more rather than less tough...it would seek to exploit the psychological advantage of its nuclear capability to intimidate the Arabs."[43] President Kennedy concurred. In a letter

---

41. Robert Rauchhaus, "Evaluating the Nuclear Peace Hypothesis," *Journal of Conflict Resolution*, Vol. 53, No. 2 (April 2009); and Michael Horowitz, "The Spread of Nuclear Weapons and International Conflict," *Journal of Conflict Resolution*, Vol. 53, No. 2 (April 2009).

42. Kapur, "India and Pakistan's Unstable Peace."

43. Central Intelligence Agency, Office of National Estimate, Memorandum for the Director, Sherman Kent, "Consequences of Israeli Acquisition of Nuclear Capability," March 6, 1963, 1, NSF, Box 118, JFKL.

to Israeli Prime Minister David Ben-Gurion, Kennedy wrote that Israel should abandon its nuclear program because nuclear proliferation in Israel would only lead to regional instability.[44] Similarly, in the case of China's nuclear program, U.S. officials believed that a nuclear-armed China would "be more willing to take risks in military probing operations because of an overoptimistic assessment of its psychological advantage."[45]

In recent years, U.S. officials have continued to fear the effect of nuclear proliferation on regional stability. In a 1986 top-secret CIA assessment, U.S. intelligence analysts predicted that a nuclear North Korea would have "a free hand to conduct paramilitary operations without provoking a response."[46] Similarly, as a U.S. expert testified before Congress in 2006, "A nuclear arsenal in the hands of Iran's current theocratic regime will be a source of both regional and global instability."[47]

U.S. officials concluded that regional instability set off by nuclear proliferation could compel them to intervene directly in regional conflicts. In the early 1960s, as chapter 3 will show, U.S. officials speculated that Israel could potentially leverage its nuclear arsenal to compel the United States to intervene on its behalf in Middle Eastern crises.[48] Similarly, in 1965, Henry Rowen, an official in the Department of Defense, postulated that if India acquired nuclear weapons, it could lead to a conflict in South Asia "with a fair chance of spreading and involving the United States."[49] Today, U.S. defense planners undoubtedly prepare for the possibility that the United States may be compelled to intervene in regional conflicts involving a nuclear-armed Iran or North Korea and their neighbors.

The empirical record has justified fears that nuclear proliferation could entangle power-projecting states in nuclear disputes. The United States has intervened in conflicts that it might have avoided, had nuclear weapons been absent. For example, the possibility that Israel might use nuclear weapons in the 1973 Yom Kippur War may have been a factor that encouraged Secretary of State Henry Kissinger to provide Israel with an emergency airlift of military equipment.[50]

---

44. John F. Kennedy, Letter to Ben Gurion, in State Department Deptel 780 (Tel Aviv), May 4, 1963, NSF, Box 119a, JFKL.
45. Burr and Richelson, "Whether to Strangle," p. 66.
46. Richelson, *Spying on the Bomb*.
47. Ilan Berman, "Confronting a Nuclear Iran," Testimony before the U.S. House of Representatives, Committee on Armed Services, February 1, 2006.
48. Cohen, *Israel and the Bomb*.
49. H. S. Rowen, "The Indian Nuclear Problem," December 24, 1964, in National Security Archive, *Nuclear Nonproliferation Policy, 1945–1990* (Alexandria, VA: Chadwyck-Healey, 1991), Document No. 01086, pp. 1, 5, as quoted in Richelson, *Spying on the Bomb*, p. 228.
50. For the role that nuclear weapons may have played in Kissinger's decision, see for example, Thomas C. Reed and Danny B. Stillman, *The Nuclear Express* (Minneapolis: Zenith Press, 2009), p. 156; and Seymour Hersh, *The Samson Option* (New York: Random House, 1991), pp. 225–240.

Similarly, in 1999 and 2002, the United States became caught in diplomatic initiatives to prevent nuclear war in crises between two nuclear powers, India and Pakistan.[51]

Indeed, the expectation that powerful states will intervene in conflicts involving a nuclear-armed state is so firmly ingrained in the strategic thinking of national leaders that small nuclear powers actually incorporate it into their strategic doctrines. South Africa's nuclear doctrine envisioned, in the event of an imminent security threat, the detonation of a nuclear weapon—not against the threatening party but over the Atlantic Ocean, in an attempt to jolt the United States into intervening on South Africa's behalf.[52] Similarly, the surprise Pakistani raid on Indian-controlled Kargil in 1999 was motivated partly by the expectation that Pakistan would be able to retain any territory it was able to seize quickly, because Pakistani officials calculated that the United States would never allow an extended conflict in nuclear South Asia.[53]

Regional instability set off by nuclear proliferation could also ensnare power-projecting states in a great power war. Other power-projecting states, facing a mirror-image situation, may feel compelled to intervene in a crisis to secure their own interests, entangling multiple great powers in a regional conflict. As we will see in chapter 3, the United States was concerned that if Israel acquired nuclear weapons, it could cause regional instability that could lead to a war between the United States and the Soviet Union. Today, U.S. strategists are concerned that China may feel compelled to intervene in any conflict involving a nuclear-armed North Korea, making the Korean Peninsula another dangerous flashpoint in the uncertain Sino-American strategic relationship.

Power-projecting states other than the United States are also threatened by the possibility that nuclear proliferation will generate regional instability that could potentially require intervention. The Soviet Union felt that nuclear proliferation in Israel could trigger regional instability that could involve Soviet forces in a broader war, as chapter 3 will show. Soviet intelligence also estimated that a South African bomb, "would lead to a sharp escalation of instability and tension in southern Africa."[54] South Korean officials similarly believe that they could become entangled in regional instability set off by nuclear proliferation in

---

For more on the U.S. role in the Yom Kippur War, see for example, Robert Dallek, *Nixon and Kissinger* (New York: HarperCollins, 2007).

51. Kapur, "India and Pakistan's Unstable Peace"; and Celia W. Dugger, "The Kashmir Brink," *New York Times*, June 20, 2002.

52. Mitchell Reiss, *Bridled Ambition* (Princeton, NJ: Princeton University Press, 1995), pp. 15–16.

53. Kapur, "India and Pakistan's Unstable Peace."

54. James Adams, *Israel and South Africa* (London: Quartet Books, 1984), p. 182.

neighboring North Korea. In the mid-1990s, Seoul prepared military forces for participation in a possible second Korean War as North Korea's nuclear program advanced.[55] Today in Turkey, strategic thinkers worry that nuclear proliferation in Iran could be a "spark [that] may be enough to 'explode' the entire region in almost every meaning of the word."[56]

Undermines Alliance Structures. During the Cold War, the Soviet nuclear threat was sometimes thought to be one of the adhesives holding the North Atlantic Treaty Organization (NATO) alliance together. The full range of effects of nuclear proliferation on alliances, however, is more complicated. Nuclear proliferation also undermines the alliance structures of power-projecting states because the spread of nuclear weapons reduces the value of the security guarantees that power-projecting states extend to their allies. Power-projecting states use the promise of military protection as a way to cement their alliance structures and to cultivate patron-client relationships. The client states are asymmetrically dependent on a relationship that ensures their survival, allowing power-projecting states influence over their clients' foreign policies. Power-projecting states can dangle—and threaten to retract—the security guarantee carrot to prevent client states from acting contrary to their interests. As nuclear weapons spread, however, alliances held together by promises of military protection are undermined in two ways. First, client states may doubt the credibility of their patron's commitments to provide a military defense against nuclear-armed states, leading them to weaken ties with their patron.[57] Second, client states can acquire nuclear weapons themselves, giving them greater security independence and making them less dependable allies. This second alliance-eroding effect is a cost that power-projecting states incur primarily when nuclear weapons spread to allied, not enemy, states.

The spread of nuclear weapons can undermine the alliance structures of power-projecting states by making allies question whether their powerful patron will be willing to come to their defense when they are threatened by a nuclear power. As Charles de Gaulle famously asked about the U.S. commitment to defend France from the Soviet Union during the Cold War, would Washington be willing to

---

55. Joel S. Wit, Daniel B. Poneman, and Robert L. Gallucci, *Going Critical* (Washington, D.C.: Brookings, 2004), pp. 176–182.
56. Kibaroglu, "Good for the Shah."
57. The deleterious effect on alliances may be exacerbated for power-projecting states with nuclear weapons because the spread of nuclear weapons may also limit their ability to provide extended nuclear deterrence to allied states, but logically, the spread of nuclear weapons threatens all power-projecting states that promise military protection to protect allies, whether the power-projecting state possesses nuclear weapons or not.

trade New York for Paris? Thomas Schelling outlined the potential steps that powerful states could take to increase the credibility of extended deterrent threats, but the sheer existence of complicated mechanisms such as "trip wires" for situations of extended deterrence do suggest that promises to defend a state against a nonnuclear weapon state are inherently more credible.[58] Accordingly, leaders in power-projecting states often worry that nuclear proliferation will undermine the credibility of their commitments, weakening the integrity of their alliance structures. John McCloy, a top advisor to the Johnson administration, argued that as nuclear weapons spread, the United States would be forced to offer security guarantees to more and more states. As McCloy put it, "The character of our determination will be diluted if we have 20 such commitments and our fundamental image of capability to defend the free world might be impaired."[59] With U.S. credibility in question, weaker allies may decide that the best way to ensure their own security would be to abandon a close security relationship with the United States. The Gilpatric Committee of 1964 speculated that if China acquired nuclear weapons, "a heightened sense of China's power could create a bandwagon effect, with greater political pressures on states in the region to accommodate Beijing and loosen ties with Washington."[60] Though the Gilpatric Committee may have overestimated this effect, the concern was not unjustified. Today, analysts point out that the development of a nuclear weapons arsenal in North Korea may already be driving a wedge between Washington and Seoul over defense policy in East Asia.[61]

Moreover, nuclear proliferation could threaten alliance cohesion by encouraging weaker allies to acquire nuclear weapons themselves. As one U.S. official pointed out during the Cold War, "European doubts about the credibility of our willingness to risk our destruction by using nuclear weapons" could "create the need for European independent capabilities."[62] If the client states themselves acquire nuclear weapons, their need for an external security guarantee is reduced, giving them greater security independence and making them less compliant to their patron's demands.[63] According to many scholars, the acquisition of the

---

58. Schelling, *Arms and Influence*.
59. William Bundy to Ambassador Thompson, "Nuclear Assurances to India," March 16, 1965, RG 59, lot 67D2, box 24, United States National Archives, College Park, Maryland (USNA), quoted in Gavin, "Blasts from the Past," p. 119.
60. Richelson, *Spying on the Bomb*, p. 144.
61. Carin Zissis, "The Fragile U.S.-South Korea Alliance," *Council on Foreign Relations*, September 2006; and "Rocket Man v. Bulldozer," *Economist*, April 5–11, 2008, p. 45.
62. "Europe, NATO, Germany, and the MLF," author and date unknown, NSF, Committee on Nuclear Proliferation, box 1, pp. 5, LBJL, quoted in Gavin, "Blasts from the Past," p. 119.
63. On nuclear weapons and security independence, see Steven Weber, "Cooperation and Interdependence," *Daedalus*, Vol. 120, No. 2 (1991), pp. 183–201.

force de frappe was instrumental in permitting the French Fifth Republic under President Charles de Gaulle to pursue a foreign policy path independent from Washington.[64]

Analysts in power-projecting states fear that the spread of nuclear weapons will shift the terms of dependence, undermining their ability to influence friendly states. As we will see in chapter 3, the United States feared that nuclear proliferation in Israel would shift the terms of the U.S.-Israeli relationship and make Washington more compliant to Israeli demands. Indeed, since Israel's acquisition of the bomb in 1967, there is no doubt that U.S. support for Israel has drastically increased. There are several reasons for the greater willingness of the United States to accommodate Israel's interests, including the strength of the pro-Israel lobby in the United States, but certainly Israel's nuclear arsenal has also increased Israel's bargaining leverage with Washington in critical moments, including perhaps, as we saw above, in the 1973 Yom Kippur War.[65] Similarly, in recent years, U.S. officials have worried about how the development of nuclear programs in places such as Taiwan and South Korea could reduce U.S. influence over its allies.[66]

The Soviet Union's threat assessments mirrored Washington's concerns about nuclear proliferation undermining alliance structures. The Soviet Union came to the conclusion, as chapter 3 will show, that nuclear proliferation in Israel would jeopardize Moscow's Middle Eastern alliances. According to Isabella Ginor and Gideon Remez, Soviet officials found that they could use their military might "to limit Israeli action against their Arab clients, thus reinforcing these clients' dependence on the USSR—as long as Israel had no counter-deterrent." Preventing Israel from acquiring nuclear weapons "thus became a central objective of Soviet Middle East policy."[67]

Dissipates Strategic Attention. The strategic consequences of nuclear proliferation listed above are reasons why power-projecting states are threatened by nuclear proliferation in and of itself. Because nuclear proliferation is so threatening to power-projecting states, however, nuclear proliferation imposes two further, secondary costs on power-projecting states. First, nuclear proliferation dissipates

---

64. See, for example, Lawrence Scheinman, *Atomic Energy Policy in France under the Fourth Republic* (Princeton, NJ: Princeton University Press, 1965); and Wilfred L. Kohl, *French Nuclear Diplomacy* (Princeton, NJ: Princeton University Press, 1971).

65. On the strength of the pro-Israel lobby in the United States, see John J. Mearsheimer and Stephen Walt, *The Israel Lobby and U.S. Foreign Policy* (New York: Farrar Straus and Giroux, 2007).

66. On U.S. concerns with nuclear programs in Taiwan, see Richelson, *Spying on the Bomb;* on U.S. concerns over South Korea, see Yoon Won-sup, "Park Sought to Develop Nuclear Weapons," *Korea Times*, January 15, 2008.

67. Ginor and Remez, *Foxbats over Dimona*, p. 32.

the strategic attention of power-projecting states. As nuclear weapons spread, power-projecting states are compelled to reapportion a costly share of strategic attention to new and potential nuclear weapon states. These burdens are shouldered by power-projecting states when nuclear weapons spread to friends and enemies alike. The collection and analysis of foreign intelligence, the practice of diplomacy, the provision of economic aid, the provision of assistance on securing new nuclear arsenals, the application of economic sanctions, defense spending, and military contingency planning focused on stopping and managing the spread of nuclear weapons—all of these require resources that will not be directed at other national goals. With each case of actual and potential nuclear proliferation within their sphere of influence, the strategic attention of power-projecting states must be spread more thinly, and the amount of resources devoted to each potential threat must be reduced.[68]

This is not to say that the strategic attention devoted to nuclear proliferation is not deserved—it is. Nuclear proliferation poses a significant threat to power-projecting states. Rather, the strategic dissipation in power-projecting states is contrasted with the position of non-power-projecting states that, as we will see below, are able to avoid expending considerable national resources on the problem of nuclear proliferation because, for them, nuclear proliferation poses less of a threat.

Strategic dissipation may seem like a trivial cost when compared to the other enumerated consequences of nuclear proliferation, but it should not be discounted. The ability of nuclear proliferation to distract the strategic attention of powerful states is an important cost of nuclear proliferation well understood by policymakers. Indeed, as we will see in subsequent chapters, states have even provided sensitive nuclear assistance with the avowed intention of distracting an important enemy.

There is also much evidence to suggest that power-projecting states devote a large portion of their strategic attention to instances of nuclear proliferation. The United States has engaged in intensive diplomacy to discourage nuclear development, drawn up military plans for possible strikes on nuclear installations, developed new military contingency plans to combat a nuclear-armed opponent, provided assistance to help the safety and security of nuclear arsenals in new nuclear states, severed economic relations with a potential nuclear proliferator in an effort to apply pressure and dissuade proliferation, and redeployed intelligence assets in order to find out more about a country's nuclear program. The United

---

68. For an assessment of the interrelationships linking the balance of power, strategic attention, and international stability, see Karl Deutsch and J. David Singer, "Multipolar Systems and International Stability," *World Politics*, Vol. 16, No. 3 (1964), pp. 390–406.

States, for example, devotes considerable resources to the problem of nuclear proliferation on the intelligence front alone. According to David Holloway:

> The United States has put an enormous effort into gathering information about the nuclear projects of other countries. After World War II it equipped aircraft with special filters to pick up radioactive debris from nuclear tests for isotopic analysis. It created a network of stations around the world to register the seismic effects of nuclear explosions. Most important, in 1960 it began to launch reconnaissance satellites that could take detailed photographs of nuclear sites in the Soviet Union and China.[69]

U.S. officials have recognized that nuclear proliferation is a considerable drain on their strategic attention. According to Francis Gavin, the Gilpatric Committee noted that in the 1960s "the U.S. Government was devoting tremendous energy to preventing other nations from acquiring" nuclear weapons.[70] Similarly, over the eight-year administration of President George W. Bush, suspected and actual nuclear weapons programs in Iran, Iraq, Libya, and North Korea, as well as the black-market nuclear exports of Pakistan, sapped an enormous amount of U.S. government resources.

Nuclear proliferation has also occupied the strategic attention of other power-projecting states. The Soviet Union redeployed intelligence assets to focus on nuclear proliferation in many countries including Israel and South Africa. It was a Soviet satellite, for example, that first detected the preparation of a nuclear test site in South Africa.[71] Moscow also drew up plans for military strikes against other states' nuclear facilities. According to one source, for example, the Soviet Union developed plans and issued orders to military commanders to strike Israel's nuclear facilities at Dimona if certain contingencies were met in the 1967 Arab-Israeli War.[72] Egypt also planned for a preventive strike on Israel's nuclear facilities.[73] Further, Egyptian President Gamal Abdel Nasser engaged in a vigorous diplomatic campaign to put international pressure on Israel's nuclear program.[74] As Pakistan marched toward the nuclear club, India redeployed intelligence assets to scrutinize Pakistan's nuclear program.[75] In recent years, South Korea has

---

69. David Holloway, "Other People's Nukes," *New York Times*, March 26, 2006.
70. Gavin, "Blasts from the Past," p. 115.
71. Peter Liberman, "The Rise and Fall of the South African Bomb," *International Security*, Vol. 26, No. 2 (Summer 2001), pp. 45–86.
72. Ginor and Remez, *Foxbats over Dimona*.
73. Ibid.
74. Cohen, *Israel and the Bomb*.
75. Perkovich, *India's Nuclear Bomb*.

expended diplomatic capital and dispensed significant economic aid in an effort to dissuade North Korea from its nuclear course.[76]

Sets Off Further Nuclear Proliferation. The other secondary cost that power-projecting states incur is that of further nuclear proliferation. Because power-projecting states are so threatened by nuclear proliferation, they frequently worry that nuclear proliferation to one state will cause further nuclear proliferation within their sphere of influence, compounding the strategic costs detailed above. When one state acquires nuclear weapons, other states may seek to develop their own nuclear arsenal in response, setting off a chain reaction of nuclear proliferation. The nuclear domino effect is probably more muted than many analysts claim, and certain policy steps, including the extension of a nuclear umbrella from a superpower, can help to mitigate the security concerns of regional states. Nevertheless, nuclear dominoes do sometimes fall. The threat of further proliferation is also a cost that is present whether nuclear weapons spread to friends or foes. After all, the acquisition of a nuclear arsenal by a close ally could encourage nuclear proliferation to other more hostile states in the same region.

Further proliferation is probably the most widely cited negative strategic consequence of nuclear proliferation recognized by analysts and policymakers in power-projecting states. For example, in 1964, U.S. Undersecretary of State George Ball predicted that a Chinese nuclear test would set off a wave of nuclear proliferation in Asia. He estimated that there was a "fifty-fifty" chance that India would follow China down the nuclear path. According to Ball, Pakistan would likely respond to India's nuclear status by seeking its own nuclear arsenal. Ball further cited Japan, Indonesia, South Korea, and Taiwan as states that could eventually develop nuclear weapons as a counter to the Chinese arsenal.[77] U.S. State Department official George McGhee also noted in 1961 that if India were to develop nuclear weapons, it could unleash "a chain reaction of similar decisions by other countries, such as Pakistan, Israel, and the United Arab Republic."[78] As we will see in chapter 3, U.S. officials also feared that Israel's nuclear program would lead to further nuclear proliferation in the Middle East. In a letter to David Ben-Gurion, President Kennedy argued that if Israel acquired nuclear weapons it would only encourage the Arab states to begin their own

---

76. Wit et al., *Going Critical*, p. 389.
77. Memcon, Couve de Murville, Charles Lucet, George Ball, and Charles Bohlen, December 2, 1964, record group (RG) 59, lot 67D2, box 7, p. 2, USNA, as cited in Gavin, "Blasts from the Past," pp. 100–135.
78. Memorandum, George McGhee to Dean Rusk, September 13, 1961, Freedom of Information Act Files, India, USNA.

nuclear weapons programs.[79] Of course, not all of the states that U.S. officials cited as potential nuclear proliferators have acquired nuclear weapons—at least not yet. Still, their fears were prescient. The Chinese bomb was a contributing cause to the development of nuclear weapons in India and, in turn, Pakistan.[80] China's nuclear arsenal was also a factor that encouraged the beginning of nuclear programs in South Korea and Taiwan. Similarly, Israel's nuclear capability sparked a nuclear program in Egypt and may have encouraged Iran's nuclear development.

In recent years, U.S. officials have stressed that nuclear proliferation in Iran and North Korea could trigger a cascade of nuclear proliferation in the Middle East and East Asia. For example, nonproliferation officials in the administration of President Bill Clinton argued that nuclear proliferation in North Korea could lead to a nuclear arms race in Asia and the potential for future nuclear weapons arsenals in South Korea, Taiwan, and Japan.[81] Similarly, in 2004, John Edwards, the Democratic Party's vice presidential nominee, stated, "A nuclear Iran is unacceptable for so many reasons, including the possibility that it creates a gateway and the need for other countries in the region to develop nuclear capability—Saudi Arabia, Egypt, potentially others."[82]

Policymakers and analysts in power-projecting states further fear that proliferation breeds proliferation by enhancing the supply of, not just the demand for, nuclear materials and technology. As the number of nuclear weapon states increases, so too does the number of states able to provide sensitive nuclear material and technology to nonnuclear weapon states, contributing to the international spread of nuclear weapons. During World War II, Selby Skinner of the U.S. Strategic Services Unit warned, "French scientists have the formula and techniques concerning atomic explosives, and ... they are now willing to sell this information ... to one of the smaller nations."[83] In the early 1990s, U.S. officials worried that South Africa could transfer HEU to other states.[84] Many analysts and policymakers have expressed concern that North Korea could export sensitive nuclear materials and technology.[85] Similarly, Peter Brookes speculated that it is possible that "Iran,

---

79. Kennedy, Letter to Ben Gurion.
80. Perkovich, *India's Nuclear Bomb*.
81. Wit et al., *Going Critical*, p. 389.
82. Glenn Kessler and Robin Wright, "Edwards Says Kerry Plans to Confront Iran on Weapons," *Washington Post*, August 30, 2004, p. A1.
83. Lt. Col. S. M. Skinner to Col. W. R. Shuler, Subject: Atomic Experiments in France, February 18, 1946, RG 226, Entry 210, Box 431, Folder 2, National Archives and Records Administration (NARA).
84. Richelson, *Spying on the Bomb*, p. 380.
85. See, for example, "President Bush's Statement on North Korea Nuclear Test."

as a nuclear weapons state, will involve itself in the dreaded 'secondary proliferation,' passing its nuclear know-how on to others."[86]

The United States is not alone in fearing that proliferation will beget proliferation; it is a concern shared by other power-projecting states. Moscow feared that nuclear proliferation in Israel would lead Moscow's Arab allies to seek nuclear weapons, as chapter 3 will show. At present, strategic thinkers in Turkey oppose nuclear proliferation in neighboring Iran because they believe that an Iranian bomb could contribute to further nuclear proliferation in their own region. Expressing the view from Ankara, Kibaroglu writes, "If Iran becomes a suspected or a *de facto* nuclear weapons state, it is feared that its neighbors such as Iraq, Saudi Arabia, Egypt, [and] Syria...may consider their nuclear options."[87]

A Consideration of Counterarguments. Some might suggest, however, that the international spread of nuclear weapons may actually be good for power-projecting states. From a theoretical standpoint, it could be argued that the provision of international security is a large burden on powerful states and nuclear proliferation could allow them to wash their hands of troubling world regions. When key states in those regions acquire nuclear weapons, according to this argument, power-projecting states will not feel the need to protect them anymore. This line of argumentation follows a clear logic, but it ignores a fundamental reality of international politics: the ability to provide for, or threaten, the security of other states is a valuable, and may even be the most important, resource in an anarchic international system. Nuclear proliferation may alleviate powerful states of a burden, but it also deprives them of much of their power and influence as well. Indeed, it is largely for this reason that power-projecting states are threatened by the proliferation of nuclear weapons.

Others may question whether these strategic effects are real and whether leaders in power-projecting states accurately assess the nuclear proliferation threat. After all, they might argue, some of the most lurid fears expressed by analysts in power-projecting states have not been borne out. For example, despite the concerns expressed by some in the Kennedy administration, nuclear proliferation in Israel did not result in a superpower war in the Middle East. While it is definitely true that many of the worst-case scenarios envisioned by analysts in power-projecting states have not come to pass, each and every one of the types of costs outlined above have been experienced by power-projecting states. In historical cases, powerful states have been deterred from using force against nuclear-armed powers; coercive threats against nuclear armed-states were never issued or

---

86. Brookes, "Iran Emboldened."
87. Kibaroglu, "Good for the Shah."

failed; powerful states have intervened in crises involving nuclear-armed states in an attempt to dial down the risk of nuclear war; alliances have frayed; strategic attention has been dissipated; and reactive proliferation has occurred. Analysts in power-projecting states, therefore, were correct in their assessment of the nature, if not always the magnitude, of the nuclear proliferation threat. In sum, leaders in power-projecting states were, and continue to be, justified in fearing the strategic consequences of nuclear proliferation.

WHY MORE IS LESS THREATENING FOR NON-POWER-PROJECTING STATES

For non-power-projecting states, the story is different. States that lack the ability to project power against a potential target state incur fewer costs when that target state acquires nuclear weapons. Non-power-projecting states lack the strategic advantages provided by conventional military power whether nuclear weapons are present or not, so nuclear proliferation does not do much to erode their strategic position. Logically, the problems that power-projecting states closely associate with the spread of nuclear weapons do not impinge upon non-power-projecting states in the same way. Again, this is true whether nuclear weapons spread to friends or enemies. Non-power-projecting states do not worry that nuclear proliferation will deter them from using military intervention to secure their interests; they are too weak to intervene militarily, whether nuclear weapons are present or not. They are not threatened by the prospect that the spread of nuclear weapons will reduce the effectiveness of their coercive diplomacy because they lack the conventional military power that could have allowed them to use threats of military force to their advantage in the first place. Non-power-projecting states do not fear that nuclear proliferation will trigger regional instability that could engulf them in conflict. Since they lack the ability to operate their military forces in and around the new nuclear weapon state, it is less likely that they could become entangled in a conflict involving the new nuclear weapon state. Non-power-projecting states need not worry that nuclear proliferation will weaken the integrity of their alliance commitments. They are too weak to promise conventional military protection as a way to cement their alliance structures. Since the strategic burdens detailed above are not borne by non-power-projecting states, they need not worry that nuclear proliferation will dissipate their strategic attention. They do not need to focus their strategic attention on nuclear proliferation because for them, these cases of nuclear proliferation are less important. Finally, non-power-projecting states are less worried that nuclear proliferation will set off further nuclear proliferation. Since they lack the ability to project power over a potential nuclear weapon state, if that state's nuclearization sends its neighbors down the nuclear path, it is also likely that the non-power-projecting state will not be able to project power over, and will not be threatened by nuclear proliferation to, the neighbors either.

This is not to say that nuclear proliferation poses no threat whatsoever to non-power-projecting states. Non-power-projecting states may still be concerned that nuclear proliferation could lead them to become the victims of a nuclear attack or nuclear coercion from a nuclear-armed state. They may fear that new nuclear states may behave more boldly in a way that could potentially be damaging to their interests. Non-power-projecting states may also worry that nuclear proliferation would deter allies from coming to their defense if they are attacked by a nuclear-armed state. Leaders in these states could also be concerned that the spread of nuclear weapons could lead to a general nuclear war among major powers. Especially in recent years, non-power-projecting states may fear that they could become the victims of nuclear terrorism.

The existence of these other potential costs of nuclear proliferation, however, does not undermine the argument that nuclear proliferation disproportionately threatens power-projecting states. First, many of these potential threats are low-probability events, making it unlikely that any state will bear these costs. After all, humanity has never experienced a nuclear exchange or a nuclear terrorist attack. By contrast, many of the constraining effects of nuclear proliferation occur with near-certainty as nuclear weapons spread. Second, it is likely that many of these additional costs are also concentrated disproportionately on power-projecting states. After all, states that have the ability to project power against a new nuclear weapon state would be more likely to come into conflict with that state. This means that power-projecting states would be more likely to be the designated target of the new nuclear state's nuclear attacks. Similarly, states with the ability to project power beyond their own borders may also be at greater risk of suffering a nuclear terrorist attack. Third, and most important for our present purposes, however, leaders in non-power-projecting states do not fear that nuclear proliferation will constrain their conventional military advantage. For this reason, the spread of nuclear weapons, on average, threatens non-power-projecting states less than it threatens power-projecting states. A summary of the differential effects of nuclear proliferation is provided in table 1.1.

This relationship between the lack of a power-projection capability and a reduction in the severity of the nuclear proliferation threat is supported by the empirical evidence. States are less threatened by nuclear proliferation to states over which they lack the ability to project conventional military power. For the most part, this evidence comes in the form of an absence of concern about the constraining effects of nuclear proliferation by strategic thinkers in non-power-projecting states. As we will see in chapters 3 and 4, the concern that nuclear

Table 1.1. The differential effects of nuclear proliferation

NUCLEAR PROLIFERATION EFFECTS	POWER-PROJECTING STATE	NON-POWER-PROJECTING STATE
Deters military intervention	Yes	No
Reduces effectiveness of military coercion	Yes	No
Triggers regional instability	Yes	No
Undermines alliance structures	Yes	No
Dissipates strategic attention	Yes	No
Sets off further nuclear proliferation within a relevant sphere of influence	Yes	No

proliferation could constrain conventional military power rarely appears in the internal strategic assessments of non-power-projecting states.

Moreover, statesmen in non-power-projecting states sometimes go further and openly proclaim that they are not threatened by nuclear proliferation. Generally, policymakers in non-power-projecting states subscribe to the idea that nuclear proliferation poses a common threat to international peace and security. Occasionally, however, leaders in non-power-projecting states are willing to state publicly that they are not threatened by nuclear proliferation. For example, as we will see in the analysis of Pakistan's exports in chapter 4, one Pakistani official openly admitted that nuclear proliferation to states beyond its own sphere of influence does not pose a direct security threat to Pakistan.

THE DIFFERENTIAL EFFECTS OF NUCLEAR PROLIFERATION AND SENSITIVE NUCLEAR ASSISTANCE

Applying this logic about the differential effects of nuclear proliferation to the specific problem of sensitive nuclear assistance suggests a number of hypotheses about the conditions under which states will provide sensitive nuclear assistance.

Power-Projection and Sensitive Nuclear Assistance. When a new state acquires nuclear weapons, the strategic costs are borne disproportionately by the states that once enjoyed the ability to project conventional military power over that state. States that are better able to project conventional military power against a particular state should fiercely oppose nuclear proliferation to that state because, in that instance, nuclear proliferation will entail a number of strategic costs. States that are less able to project military power against a particular state, on the other hand, do not incur as many strategic costs when nuclear weapons spread to that particular state, so they will be less likely to oppose nuclear proliferation to that state. We should expect, therefore, that nuclear supplier states will be reluctant to provide sensitive nuclear assistance in situations that would have the effect of imposing strategic costs on themselves. This logic gives rise to the first hypothesis:

> Hypothesis 1: The better able a state is to project power over a potential nuclear recipient state, the less likely it will be to provide sensitive nuclear assistance to that state.

It is also possible, however, that the opposite relationship between power projection and sensitive nuclear assistance holds. Power-projecting states, because they may be better able to defend against, or deter, a nuclear attack, may actually be less threatened by nuclear proliferation and may, therefore, be more likely to provide sensitive nuclear assistance. This counter-hypothesis rests on the premise, however, that capable nuclear suppliers believe that nuclear recipients would be likely to use nuclear weapons, not just to deter foreign invasion but to launch a nuclear attack. This premise is in tension with much of the nuclear deterrence literature, which maintains that nuclear weapons are useful for deterring foreign invasion but are largely ineffective instruments of warfighting.[88] Still, this is a matter that cannot be definitively settled in the theoretical realm and will be further evaluated in the empirical analysis.

Common Enemies and Sensitive Nuclear Assistance. Since nuclear proliferation entails costs for relatively powerful states, other states may actually welcome the spread of nuclear weapons in certain situations. The spread of nuclear weapons may be advantageous to states when it imposes strategic costs on other, rival states. When nuclear weapons are acquired by an enemy of an enemy, states may actually benefit because nuclear proliferation in this instance would impose particularly severe strategic costs on a rival state. Applied to the problem of sensitive nuclear assistance, we should expect that nuclear-supplier states might provide sensitive nuclear assistance in order to impose strategic costs on rivals. While nuclear proliferation to both friends and foes threatens power-projecting states, many of the costs of nuclear proliferation are most intense when nuclear weapons spread to enemy states. States could therefore provide sensitive nuclear assistance to a state with which they share a common enemy in the hope that nuclear proliferation to that state could prevent a powerful rival from intervening militarily in an area of strategic importance to the supplier state; reduce a rival's ability to use coercion as a tool of diplomacy; drag a powerful enemy into

---

88. See, for example, Richard K. Betts, *Nuclear Blackmail and Nuclear Deterrence* (Washington, D.C.: Brookings Institution Press, 1987).

regional nuclear crises; pry allies away from the force-projecting rival; reorient and dissipate a rival's strategic attention; or set off further proliferation within the sphere of influence of an enemy. For example, as we will see in chapter 3, from 1959 to 1965, France provided sensitive nuclear assistance to Israel to constrain Egypt's growing military power in the Middle East. Chapter 4 will show that Beijing's sensitive nuclear assistance to Pakistan in the early 1980s was a means of imposing strategic costs on India and diverting New Delhi's strategic attention away from China. If states are to provide sensitive nuclear assistance to constrain rival states, we should expect them to provide sensitive nuclear assistance to states with which they share a common enemy. These are precisely the situations in which a timely infusion of nuclear assistance can impose high strategic costs on a rival state. In other words, states may calculate that the enemy of their enemy is their best customer. This logic leads to the second hypothesis:

> Hypothesis 2: States will be more likely to provide sensitive nuclear assistance to states with which they share a common enemy.

The Hegemonic Nonproliferation Order and Sensitive Nuclear Assistance. States that are most disadvantaged by nuclear proliferation act to prevent it. The differential effects of nuclear proliferation invite superpower intervention. Superpowers, states with global force-projection capabilities, suffer a loss in their strategic position when nuclear proliferation occurs anywhere in the international system. For this reason, superpowers are particularly opposed to nuclear proliferation and attempt to establish a hegemonic nonproliferation order to prevent the spread of nuclear weapons.[89] It was the superpowers that led the establishment of the multilateral institutions of the nuclear nonproliferation regime, including the NPT, the Nuclear Suppliers Group (NSG), and the Zangger Committee.[90] Moreover, in individual cases of nuclear proliferation, it is often superpowers at the forefront, putting together packages of carrots and sticks to roll back the spread of nuclear weapons. Superpowers use their power and influence to dissuade other states from exporting sensitive nuclear technology.

---

89. Hegemonic stability theory focuses on the interests of powerful states in maintaining an open international economic system. The concept of a hegemonic proliferation order, in contrast, highlights a hegemon's interest in thwarting the proliferation of nuclear weapons. On hegemonic stability theory, see, for example, Stephen D. Kranser, "State Power and the Structure of International Trade," *World Politics,* Vol. 28, No. 3 (April 1976), pp. 317–347; and Robert O. Keohane, *After Hegemony* (Princeton, NJ: Princeton University Press, 1984).

90. On the creation of the nonproliferation regime, see, for example, Leonard S. Spector, *Nuclear Proliferation Today* (New York: Vintage, 1984), pp. 7–9; and William C. Potter, "U.S.-Soviet Cooperative Measures for Nonproliferation," in Rodney W. Jones, Cesare Merlini, Joseph F. Pilat, and William C. Potter, eds., *The Nuclear Suppliers and Nonproliferation* (Lexington, MA: Lexington Books, 1985).

Superpower success in inducing restraint depends on the potential supplier state's vulnerability to superpower pressure. States that depend on a superpower to meet their security needs are likely to judge that the potential costs of jeopardizing a relationship with a superpower patron outweigh the potential gains from providing sensitive nuclear assistance.[91] Superpower-dependent countries understand that any sensitive nuclear transfers would be anathema to their powerful patrons and so refrain from entering into agreements that would antagonize their security guarantors. For example, Japan is a capable nuclear supplier that depends on the United States to meet its core security needs and that has never offered to provide sensitive nuclear assistance to a nonnuclear weapon state. In the handful of historical cases in which superpower-dependent states, against their character, entered into an agreement to provide sensitive nuclear assistance, they were very likely to back down in the face of superpower displeasure. For example, as we will see in chapter 3, Argentina, a state in a formal defense pact with the United States, conceded to U.S. demands and cancelled a proposed sale of plutonium reprocessing technology to Libya in 1985.

On the other hand, states that are less dependent on a superpower patron will be more likely to provide sensitive nuclear assistance despite superpower opposition. States such as China and France, whose national nuclear weapons arsenals allow them to provide for their own security, do not rely to the same extent on superpower protection and have less to lose by antagonizing the superpowers. States that are superpower-independent will be more likely to enter into agreements to provide sensitive nuclear assistance in the first place and will be more likely to continue the transactions even in the face of superpower opposition. For example, chapter 4 will show that China, a state that was not in a formal alliance with either the United States or the Soviet Union and enjoyed the security independence afforded by a nuclear weapons arsenal, exported sensitive nuclear materials and technology to Pakistan in the 1980s, despite U.S. objections. This discussion leads us to the third hypothesis:

> Hypothesis 3: States that are dependent on a superpower patron will be less likely to provide sensitive nuclear assistance.

---

91. In theory, capable nuclear suppliers that lack nuclear weapons and that are dependent on a superpower could acquire nuclear weapons as a means of shifting the terms of dependence with the superpower. Research on the demand side of nuclear proliferation has demonstrated, however, that states in a defense pact with a nuclear power are less likely to acquire nuclear weapons. See, for example, Sonali Singh and Christopher R. Way, "The Correlates of Nuclear Proliferation," *Journal of Conflict Resolution*, Vol. 48, No. 6 (December 2004), pp. 859–885. This relationship is at least partly the result of superpower dependence as the United States and the Soviet Union attempted, and largely succeeded, to prevent their close allies from acquiring nuclear weapons. Thus, it appears that, in practice, the same superpower dependence that makes states reluctant to provide sensitive nuclear assistance also makes them reluctant to acquire nuclear weapons themselves.

These hypotheses are probabilistic and additive in nature. These factors shape the probability that sensitive nuclear transfers will occur, but they do not alone determine them. A capable nuclear supplier that lacks the ability to project power over a potential nuclear recipient, that shares a common enemy with the recipient, and that is not dependent on superpower pressure will not necessarily provide sensitive nuclear assistance and technology to that recipient, but it will be at a much greater risk of doing so. On the other hand, these hypotheses do not predict that countries that lack these characteristics will never provide sensitive nuclear assistance, but we should expect, if the strategic theory is correct, that they will be much less likely to do so.

These hypotheses are also additive. The possession of any one of the risk factors identified in the hypotheses increases the probability that a state will provide sensitive nuclear assistance, and having two or three factors raises the risk further still. The conditions identified in the hypotheses do not always align, of course, and different conditions can sometimes pull a nuclear supplier in opposing directions. For example, a state may have the ability to project power over a potential recipient but also share a common enemy with that state. The state's power projection capability would make it reluctant to provide sensitive nuclear assistance to that potential recipient, but the desire to constrain a common enemy may encourage it to do so. Still, we would expect this state to be more likely to provide sensitive nuclear assistance than a similar state that enjoyed a power-projection capability and lacked a common enemy with the recipient.

In sum, the strategic theory of nuclear proliferation presented here identifies three theoretically related, probabilistic, and additive, strategic conditions that influence the patterns of sensitive nuclear assistance.

Alternative Explanations

Economic Hypotheses

Perhaps the strongest challenge to the strategic hypotheses elucidated above is that economic incentives drive states to provide sensitive nuclear assistance. The idea that economic imperatives lead states to trade away sensitive nuclear technology is widespread. In a study on "emerging nuclear suppliers," William Potter concluded that "in addition to trying to bolster the development or, as the case may be, survival of their domestic nuclear industries, the emerging suppliers have pursued trade abroad for many of the same reasons as the traditional supplier states. Principal among these incentives is the profit motive."[92] Similarly,

---

92. William C. Potter, ed., *International Nuclear Trade and Nonproliferation* (Lexington, MA: Lexington Books, 1990), p. 412.

scholars have argued that economic underdevelopment and low levels of economic growth in North Korea could encourage Pyongyang to transfer sensitive nuclear technology in order to earn much-needed hard currency.[93] Others have argued that poor economic conditions in post-Soviet Russia may have motivated civilian nuclear exports to Iran.[94] For example, a recent analysis of Russian assistance to Iran's nuclear program argued that "transfers of nuclear technology from Russia to Iran...cannot be ruled out given the financial crisis in the Russian nuclear industry."[95] This logic suggests two hypotheses. First, less-developed states may be more likely to provide sensitive nuclear assistance. Thus, among potential suppliers, lower levels of economic development may be associated with a higher propensity to export sensitive nuclear technology and materials. Second, states experiencing low rates of economic growth may be more likely to take extreme measures, such as the export of sensitive nuclear technology and materials, to improve their economic circumstances.

There are other potential economic explanations for the patterns of sensitive nuclear assistance. Jabko and Weber have argued that because of its dependence on international trade, France may be more likely than other states to export civilian nuclear technologies.[96] We may expect, therefore, that states that are more open to the international economy will be more likely to provide sensitive nuclear assistance. Similarly, states that are dependent on a particular trading partner may be more likely to provide sensitive nuclear assistance to that state to avoid undermining an important trade relationship.

There are also reasons, however, to be skeptical about the role of economics in explaining the transfer of sensitive nuclear technology and materials. The direct economic rewards for exporting sensitive nuclear fuel-cycle facilities are significantly lower than those associated with the provision of civilian nuclear assistance.[97] While a contract for the construction of a nuclear power reactor can be quite lucrative, the reward for transferring most sensitive fuel-cycle technologies, such as a plutonium reprocessing lab, is not. For example, in 1975, the price for a plutonium reprocessing laboratory was estimated at a mere 1.6 *million* dollars.[98]

---

93. See, for example, Sheena Chestnut, "Illicit Activity and Proliferation," *International Security*, Vol. 32, No. 1 (Summer 2007), pp. 80–111.
94. Vladimir A. Orlov and Alexander Vinnikov, "The Great Guessing Game," *Washington Quarterly*, Vol. 28, No. 2 (Spring 2005), pp. 49–66.
95. Nuclear Threat Initiative, *Russia Nuclear Exports to Iran: Reactors*.
96. Nicholas Jabko and Steven Weber, "A Certain Idea of Nuclear Weapons: France's Non-Proliferation Policies in Theoretical Perspective," *Security Studies*, Vol. 8, No. 1 (Fall 1998), pp. 108–150.
97. William C. Potter, *Nuclear Power and Nonproliferation* (Cambridge, MA: Oelgeschlager, Gunn, and Hain, 1982).
98. Steve Weissman and Herbert Krosney, *The Islamic Bomb* (New York: New York Times Books, 1981), p. 98.

In the same year, the price for a nuclear power reactor was estimated at between two and eight *billion* dollars.[99] In fact, the aforementioned study by William Potter also found, in direct tension with its primary conclusion, that in practice "nuclear exports generally have not yielded substantial economic returns for the emerging suppliers."[100]

Moreover, states with nuclear fuel-cycle capabilities may reap greater economic gains by providing fuel-cycle services to other states, rather than by exporting the fuel-cycle facilities themselves.[101] The export of capabilities to other states shrinks the customer base for fuel-cycle services and creates potential competitors in the market for the supply of fuel-cycle services. Indeed, it was this that led Potter to argue that for some nuclear suppliers, "the economic interests of their nuclear industries would not appear to suffer substantially from a nonproliferation strategy which restricted nuclear exports."[102]

Not only are the direct economic rewards to sensitive nuclear assistance slight, a broader economic relationship may also be undermined by sensitive nuclear transfers. Etel Solingen and T. V. Paul have argued that states that are open to the international economy may be less likely to seek nuclear weapons because they are reluctant to risk international trade and investment on controversial foreign policies.[103] Correspondingly, states that engage in international trade may be less likely to provide sensitive nuclear assistance to avoid jeopardizing important international economic relationships.

While there are theoretical reasons to be skeptical about the importance of economic factors for explaining sensitive nuclear assistance, the economic hypotheses elucidated above will be further evaluated against the empirical evidence in the remaining chapters of this book.

Institutional Hypotheses

International institutions are thought to affect state behavior through the establishment of formal rules that regulate state action and of informal norms that shape understandings of appropriate conduct.[104] The institutions of the nuclear

---

99. Norman Gall, "Atoms for Brazil, Dangers for All," *Foreign Policy*, Vol. 23 (Summer 1976), pp. 155–201.
100. Potter, *International Nuclear Trade*, p. 412.
101. Potter, *Nuclear Power and Nonproliferation*, p. 106.
102. Ibid., p. 106.
103. Etel Solingen, *Nuclear Logics* (Princeton, NJ: Princeton University Press, 2007); and T. V. Paul, *Power Versus Prudence* (Montreal: McGill-Queen's University Press, 2000).
104. On international institutions, see, for example, Stephen Krasner, ed., *International Regimes* (Ithaca, NY: Cornell University Press, 1983); and Barbara Koremenos, Charles Lipson, and Duncan

nonproliferation regime set restrictions on the transfer of nuclear materials and technology, which may render member states less likely to provide sensitive nuclear assistance. According to Article II of the NPT, nonnuclear weapon states agree "not to seek or receive any assistance in the manufacture of nuclear weapons or other nuclear explosive devices." And Article I binds each nuclear weapon state "not in any way to assist, encourage, or induce any non-nuclear weapon state to manufacture or otherwise acquire nuclear weapons or other nuclear explosive devices, or control over such weapons or explosive devices."[105] The NSG, a cartel of capable nuclear suppliers established in 1975, adds further restrictions on the export of sensitive nuclear technology. NSG members agree to "exercise restraint in the transfer of sensitive facilities, technology and weapons-usable materials."[106] The formal rules enshrined in the international institutions of the nuclear nonproliferation regime may discourage member states from transferring sensitive nuclear materials and technology. The establishment of the NPT may have also led to the creation of informal norms, proscribing the proliferation of nuclear weapons. For these reasons, we may expect that members of the NPT and the NSG will be less likely to provide sensitive nuclear assistance. It is also possible that normative effects of the nuclear nonproliferation regime may exert some pressure on all states in the international system whether or not they are members. This logic may lead us to believe that since the creation of the NPT, all states, member and nonmember states alike, will be less likely to provide sensitive nuclear assistance. Or it may be that certain types of states that are predisposed to abiding by international norms, such as states that are governed domestically by democratic institutions, will be less likely to provide sensitive nuclear assistance.

Skeptics argue, however, that international institutions may not pose meaningful constraints on state behavior.[107] To paraphrase Richard Betts, when international institutions work they are not needed, and when they are needed they do not work.[108] States may simply refuse to adhere to their treaty commitments when it is in their interest to do so. Further, if states that are members of international institutions behave differently than nonmembers, it may simply be the

---

Snidal, "The Rational Design of International Institutions," *International Organization*, Vol. 55, No. 4 (Autumn 2001), pp. 761–799.
    105. Treaty on the Nonproliferation of Nuclear Weapons, 1968.
    106. Nuclear Suppliers Group, "Guidelines for Nuclear Transfers."
    107. For a critical view on the effectiveness of international institutions, see John J. Mearsheimer, "The False Promise of International Institutions," *International Security*, Vol. 19, No. 3 (Winter 1994/1995), pp. 5–49; and Richard K. Betts, "Systems for Peace or Causes of War?" *International Security*, Vol. 17, No. 1 (Summer 1992), pp. 5–43.
    108. Betts, "Systems for Peace or Causes of War?"

result of a screening effect.[109] This means, in short, that states that decide to join international institutions are different from states that decide not to join. It may be that membership in the nuclear nonproliferation regime does not make one less likely to provide sensitive nuclear assistance, but rather that states with less inclination to provide sensitive nuclear assistance are more likely to join the institutions of the nuclear nonproliferation regime.

In sum, there are reasons to believe that international institutions may prevent sensitive nuclear assistance alongside other reasons to believe that institutions may exert no meaningful effect on sensitive nuclear transfers. These contrasting theoretical expectations will be examined further in the empirical component of this book.

Nuclear Weapons Possession Hypothesis

Some scholars have argued that nuclear weapon states should be more likely than nonnuclear weapon states to oppose nuclear proliferation to additional states because nuclear weapon states have strategic incentives to maintain an exclusive nuclear club. George Quester, for example, has claimed that a nuclear weapon state has "an interest in shutting the [nuclear] door behind itself."[110] Joseph Pilat has similarly argued that "France, as a nuclear-weapon state, does have...a strategic interest in [non] proliferation."[111] Jonathan Pollack concurs with this logic, arguing that China, as a member of the nuclear club, would "want it to be a small club."[112] Applying this insight to the problem of sensitive nuclear assistance, we may hypothesize that nuclear weapon states will be less likely to provide sensitive nuclear assistance than capable nuclear suppliers that lack nuclear weapons themselves.

Note that this hypothesis is different from the argument that power-projecting states want to preserve their conventional military freedom of action. Rather, this argument is about nuclear weapon states wanting to contain the spread of nuclear weapons in order to limit any security or prestige-related benefits related

---

109. For a recent manifestation of the debate over whether international agreements constrain or screen for state behavior, see Jana Von Stein, "Do Treaties Constrain or Screen?" *American Political Science Review*, Vol. 99, No. 4 (November 2005), pp. 611–623; and Beth A. Simmons and Daniel J. Hopkins, "The Constraining Power of International Treaties," *American Political Science Review*, Vol. 99, No. 4 (November 2005), pp. 623–633.

110. George Quester, "The Statistical 'N' of the 'Nth' Nuclear-Weapon States," *Journal of Conflict Resolution*, Vol. 27, No. 1 (March 1983), p. 175.

111. Joseph F. Pilat, "The French, Germans, and Japanese and the Future of the Nuclear Supply Regime," in Jones et al., *Nuclear Suppliers and Nonproliferation*, p. 82.

112. As cited in Rone Tempest, "Dangerous Dynamic between China and India," *Los Angeles Times*, June 13, 1998.

to nuclear weapons possession itself to as few other states as possible. Given that there are many states that have nuclear weapons but lack global force projection capabilities, such as Pakistan, and many other states, for example Egypt, that have the ability to project conventional military power over states in their region but lack nuclear weapons, these hypotheses generate divergent and testable hypotheses about the determinants of sensitive nuclear assistance. If the nuclear club argument is correct, we should expect nuclear weapon states, whether or not they possess power-projection capabilities, to be extremely reluctant to provide sensitive nuclear assistance.

We may also expect the exact opposite relationship, however, between nuclear weapons possession and the probability of sensitive nuclear assistance. Countries with a nuclear deterrent can better provide for their own security and should be less vulnerable to superpower pressure. Nuclear weapon states, because they are relatively insulated from superpower pressure, may be more likely to export sensitive nuclear technology and materials. It is also possible that nuclear weapon states may be more likely to provide sensitive nuclear assistance because the possession of nuclear weapons gives them confidence that they could deter attacks from new nuclear weapon states. Furthermore, nuclear weapon states may be more prolific nuclear suppliers because they are better able to provide nuclear assistance related to the design and construction of nuclear weapons.

In sum, it is possible that nuclear weapon states may be less (or more) likely to provide sensitive nuclear assistance than capable nuclear suppliers that lack nuclear weapons. The relationship between nuclear weapons possession and sensitive nuclear assistance will be explored in the empirical chapters to follow.

Alliance Hypothesis

One could argue that states provide sensitive nuclear assistance in order to strengthen allied states. Joanne Gowa has argued that states benefit from strong allies and that states engage in behavior that empowers allied states.[113] One might also argue that states could transfer nuclear weapons technologies and materials to allied states in order to avoid the extended deterrence problems associated with supplying a security umbrella or as a way to gain increased influence over an ally's foreign policy. We may expect, therefore, that states will be more likely to transfer sensitive nuclear materials and technology to allied states.

---

113. Joanne Gowa, *Allies, Adversaries, and International Trade* (Princeton, NJ: Princeton University Press, 1995).

There are also, however, theoretical and empirical reasons to be very skeptical of alliance-based arguments for sensitive nuclear transfer. As argued above, nuclear-armed states can be unreliable allies. They have the autonomy that allows them to forge an independent security policy.[114] Providing sensitive nuclear assistance to an allied state may be a better means of losing an ally than of creating a strong and reliable partner. Indeed, a cursory examination of the evidence reveals that this hypothesis cannot generate enough empirical support to warrant inclusion here. Every single case of sensitive nuclear assistance has occurred between states that lack a formal alliance.[115] Not only are states not more likely to provide sensitive nuclear assistance to an allied state, there is no recorded instance of such an act.[116] This may suggest that states transfer sensitive nuclear technology and materials not according to the state they intend to help but, consistent with Hypothesis 2, according to the state they intend to constrain.[117]

Research Design and Case Selection

To adjudicate among the strategic, economic, institutional, and nuclear possession hypotheses presented in this chapter, I will use both quantitative and qualitative methodologies. The large-N analysis will draw on an original sensitive nuclear assistance dataset to analyze the determinants of sensitive nuclear assistance. The design of the quantitative study will be explained in further detail in chapter 2. In the remainder of this section, I will discuss the selection of the qualitative case studies that will be featured in chapters 3 and 4.

Methodologists often advocate selecting cases on the independent variable to avoid the selection bias that is inherent in studies that select on the dependent variable.[118] When studying rare-event phenomena, however, selecting on the independent variable can be problematic because researchers run the risk of

---

114. Weber, *Cooperation and Interdependence*, 1991.
115. The Correlates of War alliance variable measures three types of alliance relationship—defense pact, entente, neutrality agreement—in addition to no alliance. Of the seventy-nine dyad years of international nuclear assistance, all seventy-nine are between states in the "no alliance" category.
116. For a list of cases of sensitive nuclear assistance, see appendix C. For a list of cases that do not qualify as sensitive nuclear assistance, see appendices D and E.
117. The evidence not only undermines the ally-strengthening argument for international nuclear assistance, it also challenges the positive externalities of trade literature. Gowa argues that states act in order to strengthen allies. There is no more direct way of strengthening an ally than a large-scale transfer of military technology, yet states resist from transferring sensitive nuclear materials and technology to allied states.
118. Gary King, Robert O. Keohane, and Sidney Verba, *Designing Social Inquiry* (Princeton, NJ: Princeton University Press, 1994), pp. 129–132; and Barbara Geddes, "How the Cases You Choose Affect the Answers You Get," *Political Analysis*, Vol. 2, No. 1 (1990), pp. 131–150.

choosing cases that provide no variation on the dependent variable. For example, if in this study on sensitive nuclear assistance, I selected cases on the independent variable alone, I would likely select a number of cases in which sensitive nuclear assistance never occurred. If a study lacks variation on independent and dependent variables, however, the analyst cannot explore causal relationships among variables. Moreover, it is widely observed that, in practice, scholars often have foreknowledge of the outcome of the cases before critical case selection decisions are made. For both of these reasons, methodologists also recommend selecting cases to ensure variation on both the independent and dependent variables.[119] Crossing high and low levels of a key independent variable with high and low levels of the dependent variable creates a two-by-two matrix. Selecting at least one case from each quadrant provides the analyst with cases that are congruent with the hypothesized relationship between the independent and dependent variables. It also ensures the selection of outliers. These will be cases in which independent and dependent variables are incongruent with the theory's predictions, allowing the analyst to test the limits of the theory.

As we can see in table 1.2, this study analyzes cases in which there is variation on a key independent variable (power-projection capability) and the dependent variable (sensitive nuclear assistance). French assistance to Israel from 1959 to

Table 1.2. Case selection

POWER-PROJECTING STATE	SENSITIVE NUCLEAR ASSISTANCE	
	YES	NO
Yes	U.S.S.R.: China (1958–1960) chapter 4	U.S.: Israel (1959–1965) chapter 3
No	France: Israel (1959–1965) chapter 3	Israel: various (1966–present) chapter 4
	China: Pakistan (1981–1986) chapter 4	India: various (1964–present) chapter 4
	Pakistan: various (1987–2002) chapter 4	

---

119. King et al., *Designing Social Inquiry*, pp. 142–146.

1965 is a prime example of a state that provided sensitive nuclear assistance to a state over which it lacked an effective power-projection capability. This case also provides the researcher with historical perspective, declassified primary materials, and abundant secondary analysis. Thus, the French-Israel case will receive a thorough treatment in chapter 3. Other important cases that occupy this quadrant, including China's exports to Pakistan and Pakistan's nuclear transfers to Iran, Libya, and North Korea, will be examined in chapter 4.

While there are a vast number of cases in which powerful states refrain from providing sensitive nuclear assistance to other states, pairing the France-Israel case with an analysis of the potential role of the United States as a nuclear supplier to Israel from 1959 to 1965 helps to control for other factors that could influence the analysis. Both France and the United States experienced similar economic situations in this time period. They were both advanced industrialized countries with rapidly growing economies. Their memberships in the relevant institutions of the nuclear nonproliferation regime were also similar. Given that France's transfers to Israel occurred before the creation of the nuclear nonproliferation regime, neither state was a formal member of any relevant institutions. This "most-similar-systems" design allows for a structured, focused comparison of the way in which a power-projecting state, the United States, and a non-power-projecting state, France, viewed Israel's nuclear ascendancy.[120] The outcomes of these two cases are congruent with the predictions of the theory and will form the content of chapter 3.

Selecting a case of a power-projecting state that provided sensitive nuclear assistance is straightforward. Soviet nuclear assistance to China (1958–1960), a case of a superpower transferring nuclear technology to a state over which it could have easily projected military power, cries out to be studied. Israel and India are also attractive candidates for a study of non-power-projecting states that refrain from providing sensitive nuclear assistance. Along with Japan, Israel and India are the only states that fit this description over the entire course of the nuclear era.[121] While Japan's restraint is easily explained by Japan's dependence on the U.S. nuclear umbrella, the lack of sensitive nuclear exports originating from Israel and India is more puzzling. Furthermore, a study of Israel as a supplier allows the analysis of at least one state as both a potential recipient and

---

120. On most-similar-systems design, see Adam Przeworski and Henry Teune, *Logic of Comparative Social Inquiry* (New York: Krieger, 1982). On the method of structured, focused comparison, see Alexander L. George, "Case Studies and Theory Development," in Paul Gordon Lauren, ed., *Diplomacy* (New York: Free Press, 1979), pp. 43–68.

121. Other small powers that have traditionally adopted tough anti-nuclear policies, including Canada, Ireland, New Zealand, Sweden, and Switzerland, lack the ability to provide sensitive nuclear assistance. See table 2.1 for a list of capable nuclear suppliers.

a potential supplier, allowing the researcher to control for other country-specific factors that could influence a state's nonproliferation behavior. Soviet assistance to China and Israel and India's behavior as potential nuclear suppliers are inconsistent with the predictions of the theory and will help to test the theory's limits. These cases will be examined in chapter 4. In both chapters, I will also draw on a number of other short shadow cases as appropriate in order to improve causal inference.[122]

We will return to examine the details of these cases in future chapters. At this point, however, we will put the discussion of the qualitative case studies on hold. The next chapter will present the results of a quantitative analysis of the correlates of sensitive nuclear assistance.

---

122. For a discussion of the value of adding cases to one's study to improve causal inference, see King, Keohane, and Verba, *Designing Social Inquiry*, p. 219.

2

THE CORRELATES OF NUCLEAR ASSISTANCE

This chapter presents a number of statistical tests, drawing on an original sensitive nuclear assistance dataset, to analyze the determinants of nuclear assistance. The analysis provides strong support for the strategic theory of nuclear proliferation presented in this book. In accordance with Hypothesis 1, the less able a state is to project power over a potential nuclear recipient, the more likely it is to provide sensitive nuclear assistance to that state. States are more likely to export sensitive nuclear materials and technology to states with which they share a common enemy, consistent with Hypothesis 2. And, in support of Hypothesis 3, states that are less vulnerable to superpower pressure are more likely to conduct sensitive nuclear transfers. Among the major alternative explanations for sensitive nuclear assistance, the institutions argument is the only one that seems to bear on the probability of sensitive nuclear assistance. This chapter finds some evidence for the idea that the NPT constrains member states from exporting sensitive nuclear materials and technology. In what may come as a surprise to some, economic factors are unrelated to sensitive nuclear assistance.

Sensitive Nuclear Assistance Data

To test the strategic theory of nuclear proliferation, I construct an original sensitive nuclear assistance dataset. The dataset contains yearly information for all capable nuclear suppliers and potential nuclear recipients in the international system from 1951 to 2000.[1] Capable nuclear suppliers are states that could conceivably transfer sensitive nuclear materials and technology to other states. States are considered to be capable nuclear suppliers beginning in the first full year in which they successfully operate a domestic plutonium reprocessing or uranium enrichment facility. This group of states includes nuclear powers such as France, Pakistan, and the United States as well as states such as Brazil, Germany, and Japan that have mastered parts of the nuclear-fuel cycle but do not possess nuclear weapons themselves. A list of capable nuclear suppliers is presented in Table 2.1. Potential nuclear recipients are all nonnuclear weapon states in the international system. States with nuclear weapons are not included as potential recipients because the puzzle motivating this study concerns the motivations leading states to provide sensitive nuclear assistance to nonnuclear weapon states.

The dichotomous dependent variable is *Sensitive nuclear assistance*.[2] It measures whether a capable supplier state provided sensitive nuclear assistance to a potential nuclear recipient in a given year.[3] Appendix C lists the cases of sensitive nuclear assistance and explains the key coding decisions.[4] A selection of cases that do not qualify as sensitive nuclear assistance according to the above definition can be found in appendices D and E.

I construct independent variables to test the strategic hypotheses explicated above. *Power projection*, a supplier state's ability to project power over a potential recipient, is measured as the military capability of the supplier state, discounted by distance between the supplier and the potential recipient state, minus the military capability of the potential recipient state. To measure military capability, I employ the most commonly used measure of military power in the field of international relations. The Composite Index of National Capability (CINC) score, is a composite index containing information on total population, urban population, energy consumption, iron and steel production, military manpower,

---

1. The unit of analysis is the directed-dyad year.
2. To code the sensitive nuclear assistance variable, I began with an online nuclear weapons database maintained by the Nuclear Threat Initiative. I also drew on prominent reviews of the proliferation of nuclear weapons and on historical studies of countries' nuclear weapons programs. To be included in the dataset, a case of sensitive nuclear transfer had to be verified by at least two sources.
3. I code as ones all years in which sensitive nuclear assistance occurs, because my theoretical interest is in the occurrence of sensitive nuclear assistance, not just its onset. Using an onset of nuclear assistance variable produces similar results.
4. The secretive nature of sensitive nuclear transfers raises the possibility of a missing data problem. There may be sensitive nuclear transfers that have occurred that we do not yet know about. It is difficult, however, for countries to maintain a secret nuclear program, and sensitive nuclear transfers that countries attempt to conduct in secret generally become known within a few years. Given the time frame of this study, which ends in 2000, I assess that missing data does not pose a significant problem to this analysis.

Table 2.1. Capable nuclear suppliers

COUNTRY	YEAR	NOTES
Argentina	1969	Begins operation of hot cells at Ezeiza. Spector, *Nuclear Proliferation Today*, 1984, p. 203.
Belgium	1966	Eurochemic center for reprocessing technology begins operation. *The Decommissioning of the Eurochemic Reprocessing Plant*, www.belgoprocess.be/03_act/docs/BP02_Eurochemic.pdf.
Brazil	1988	Inaugurates enrichment facilities. Albright, Berkhout, and Walker, *Plutonium and Highly Enriched Uranium 1996*, pp. 374–75.
China	1964	Uranium enrichment begins. Lewis and Litai, *China Builds the Bomb*, pp. 136, 186–189.
France	1958	First reprocessing plant starts at Marcoule. World Nuclear Association, *Country Briefings: France*.
Germany	1971	As part of URENCO consortium, begins operating enrichment facilities. URENCO, *URENCO Company History*.
India	1964	Begins plutonium reprocessing. Jones et al., *Tracking Nuclear Proliferation*, pp. 112, 129.
Israel	1966	Begins plutonium reprocessing. Cohen, *Israel and the Bomb*, p. 231.
Italy	1970	Begins plutonium reprocessing. Nuclear Energy Agency, *Decommissioning of Nuclear Installations in Italy (Jan 2006)*.
Japan	1977	First plutonium reprocessing facility goes online. Reiss, *Without the Bomb*, p. 113.
The Netherlands	1971	As part of URENCO consortium, begins operating enrichment facilities. URENCO, *URENCO Company History*.
North Korea	1993	Begins plutonium reprocessing. Jones et al., *Tracking Nuclear Proliferation*, p. 147.
Norway	1961	Pilot-scale reprocessing begins. Nuclear Energy Agency, *Decommissioning in NEA Member Countries, Current Status: Norway*.
Pakistan	1986	Begins uranium enrichment. Jones et al., *Tracking Nuclear Proliferation*, p. 132.
USSR/Russia	1949	Begins enrichment and reprocessing. Holloway, *Stalin and the Bomb*, p. 188.
South Africa	1977	Uranium enrichment begins in South Africa. Jones et al., *Tracking Nuclear Proliferation*, pp. 243–244.
United Kingdom	1951	Begins plutonium reprocessing domestically. Gowing, *Independence and Deterrence*.
United States	1945	This study begins at the dawn of the nuclear era, when the United States used nuclear weapons against Japan in 1945.
Yugoslavia/Serbia	1966	Pilot-scale plutonium reprocessing begins. Nichol and McDaniel, "Yugoslavia," in Katz and Marwah, eds., *Nuclear Power in Developing Countries*.

and military expenditures.⁵ A state's ability to project power over another state depends on the geographical distance between the two states. Militaries, like baseball teams, are more effective when they enjoy a home field advantage.⁶ For this reason, I discount the capabilities of the supplier by the distance between itself and the recipient using a formula advanced by Bruce Bueno de Mesquita.⁷ I also construct an alternative *Power ratio* variable, measured as the composite capability of the supplier state divided by the combined composite capability of the supplier and the recipient.

To measure whether the nuclear supplier and the nuclear recipient are threatened by a common rival, I construct an *Enemy* variable. This dichotomous variable indicates whether the nuclear supplier and the nuclear recipient share a common rival. The variable is based on a definition of rivalry from Klein, Goertz, and Diehl that measures a rivalry as a pair of states that engage in multiple and linked, militarized interstate disputes within a given time period.⁸

I construct three variables to measure a state's vulnerability to superpower pressure. *Superpower pact* is a dummy variable, indicating whether a state relies on a superpower security guarantee.⁹ States that rely on a superpower security guarantee are states that lack a nuclear weapons capability and are in a defense pact with a superpower.¹⁰ The United States, from 1951 to 2000, and the Soviet Union, from 1951 to 1989, are coded as superpowers. An alternate measure, *Superpower vote*, measures the similarity of states' voting behavior in the United Nations General Assembly (UNGA) with the voting behavior of the nearest superpower.¹¹ A state that is vulnerable to superpower pressure is likely to vote with,

---

5. Data on capabilities are drawn from the Correlates of War composite capabilities index, version 3.1 and extracted using EUGene. On the composite capabilities index, see J. David Singer, Stuart Bremer, and John Stuckey, "Capability, Distribution, Uncertainty, and Major Power War, 1820–1965," in Bruce Russett, ed., *Peace, War, and Numbers* (Beverly Hills, CA: Sage, 1972). On EUGene, see Scott D. Bennett and Allan Stam, "*EUGene*: A Conceptual Manual," *International Interactions*, Vol. 26, No. 2 (2000), pp. 179–204.

6. For a seminal piece on geographic distance and military capability, see Kenneth Boulding, *Conflict and Defense* (New York: Harper and Brothers, 1962).

7. Bruce Bueno de Mesquita, *The War Trap* (New Haven: Yale University Press, 1981).

8. James P. Klein, Gary Goertz, and Paul F. Diehl, "The New Rivalry Dataset," *Journal of Peace Research*, Vol. 43, No. 3 (May 2006), pp. 331–348.

9. Data on defense pacts and nuclear weapons possession are drawn from version 3.0 of the Correlates of War alliance data set and on Gartzke and Kroenig's coding of nuclear weapon states, respectively. On the Correlates of War alliance data set, see Douglas M. Gibler and Meredith Sarkees, *Coding Manual for v3.0 of the Correlates of War Formal Interstate Alliance Data Set, 1816–2000*, 2002. For the coding of nuclear weapon states, see Gartzke and Kroenig, "Strategic Approach to Nuclear Proliferation."

10. Following Singh and Way, "Correlates of Nuclear Proliferation," I only count defense pacts as providing a security guarantee.

11. Data on UNGA voting are drawn form Erik Gartzke, *Codebook for the Affinity of Nations Index, 1946–2002*, version 3.0, 2006.

rather than against, its patron. From 1951 to 1989, this variable measures the similarity of states' voting behavior with the superpower that has the most similar voting profile in each year. From 1990 to 2000, this variable measures the similarity of states' voting behavior with the United States. For a final, alternate measure of superpower dependence, I code *Nuclear weapon,* which measures whether a country has a nuclear weapons arsenal.[12] Countries with a nuclear deterrent can better provide for their own security and should be less vulnerable to superpower pressure. The inclusion of this variable also permits an evaluation of the alternative hypothesis that nuclear weapon states will be less likely to provide sensitive nuclear assistance because they have an interest in limiting the size of the nuclear club.

I also include a number of variables to test alternative arguments and to control for other factors that could influence the patterns of sensitive nuclear assistance. For more detail on each, see the discussion in appendix A. To assess the economic arguments, I include variables that gauge the economic circumstances of the nuclear supplier, including level of economic development, economic growth, dependence on international trade, and dependence on trade with potential nuclear recipients. To test the effects of international institutions on sensitive nuclear assistance, I code whether states were members of the key institutions of the nuclear nonproliferation regime. I control for factors that influence the recipient states' demand for nuclear assistance, including level of economic development, openness to the international economy, economic liberalization, threat environment, and dependence on a superpower patron. I also control for the geographical distance between the supplier and recipient states and the regime type of the supplier.

Data Analysis

Sensitive nuclear assistance, like wars, revolutions, and other important events in international politics, is a rare event. Sensitive nuclear assistance occurs in about one out of every one thousand possible cases.[13] To study these important and rare events, scholars have developed specialized statistical techniques. I employ Rare Events Logistic Regression (ReLogit) to test claims about the correlates of sensitive nuclear assistance.[14] ReLogit offers several advantages of particular

---

12. Data on nuclear weapons possession are drawn from Gartzke and Kroenig, "Strategic Approach to Nuclear Proliferation."
13. More precisely, sensitive nuclear assistance occurs in 79 of the 81,952 dyad-year observations.
14. On ReLogit, see Gary King and Langche Zeng, "Logistic Regression in Rare Events Data," *Political Analysis,* Vol. 9, No. 2 (Spring 2001), pp. 137–163.

relevance to this research question and data.[15] ReLogit is able to model dichotomous dependent variables and to correct for biased estimates in rare events. In particular, ReLogit is designed to analyze "binary dependent variables with dozens to thousands of times fewer" events than nonevents.[16]

To begin the statistical investigation, I examine the simple bivariate relationship linking the key strategic, economic, and institutional variables and sensitive nuclear assistance.[17] For each measure, I also examine the bivariate relationship after the inclusion of cubic splines to control for temporal dependence in the dependent variable.[18] The bivariate analysis is presented in Table 2.2. The bivariate analysis is only the first step, however. To control for potentially confounding factors, I then evaluate the effect of each of the explanatory variables, including both control variables and the cubic splines. The results of the multivariate analysis are displayed in Table 2.3. I also performed a battery of tests to assess the robustness of the statistical findings. The robustness checks are discussed in appendix A.

I first examine the hypothesis that *Power projection* is negatively related to sensitive nuclear assistance. Hypothesis 1 states that the better able a state is to project power over a potential nuclear recipient, the less likely it will be to provide sensitive nuclear assistance to that state. Turning first to the bivariate models, we see that the relationship between power projection and sensitive nuclear assistance is negative and statistically significant in each and every model. Next, an examination of the multivariate regressions reveals a similar pattern. Again, the sign on the coefficient is negative and statistically significant in each model.[19] There is strong empirical support for the causal significance of the supplier's ability to project power over the recipient for understanding sensitive nuclear assistance. The counter-hypothesis that power-projecting states will be more likely to provide sensitive nuclear assistance because they can better deter a nuclear attack does not find empirical support.

---

15. Using other statistical estimators such as Logit or Complementary Log Log instead of ReLogit did not alter the statistical significance or the direction of the sign on the coefficients of the results reported below.

16. King and Zeng, "Logistic Regression in Rare Events Data," p. 137.

17. I examine each supplier-recipient dyad in each year, but not all of these observations are statistically independent of each other. For example, China's relationship with Pakistan in 1983 is not independent from the relationship between the same two countries in the subsequent year. The presence of such dependent observations could deflate the standard errors. To address this problem, I corrected the standard errors for clustering.

18. Nathaniel Beck, Jonathan N. Katz, and Richard Tucker, "Taking Time Seriously in Binary Time-Series-Cross-Section Analysis," *American Journal of Political Science*, Vol. 42, No. 4 (October 1998), pp. 1260–1288.

19. These results were robust when *Power ratio* was substituted for *Relative power*.

Table 2.2. Correlates of sensitive nuclear assistance, 1951–2000: Bivariate results

INDEPENDENT VARIABLE	COEFFICIENT	ROBUST STANDARD ERROR	N	MODEL
STRATEGIC VARIABLES				
H1: Power projection	−53.371	3.396***	81952	ReL
	−49.037	7.772***	81952	CS
H2: Enemy	2.376	0.234***	81952	ReL
	2.545	0.547***	81952	CS
H3: Superpower pact	−0.961	0.248***	81952	ReL
	−1.033	0.662	81952	CS
H3: Superpower vote[1]	−0.384	0.275	81127	ReL
	−1.160	0.399**	81127	CS
H3: Nuclear weapon	0.752	0.242**	81952	ReL
	1.192	0.606*	81952	CS
ECONOMIC VARIABLES				
A1: Economic development	−0.065	0.018***	81952	ReL
	−0.002	0.063	81952	CS
A2: Growth	5.239	3.011	81952	ReL
	5.503	4.044	81952	CS
A3: Openness	−0.007	0.002***	81952	ReL
	−0.001	0.005	81952	CS
A4: Trade dependence	12.020	1.886***	81952	ReL
	16.872	4.361***	81952	CS
INSTITUTIONAL VARIABLES				
A5: NPT	−1.137	0.2477***	81952	ReL
	−0.399	0.695	81952	CS
A6: NSG	−0.400	0.237	81952	ReL
	0.673	0.659	81952	CS

Notes: Each row in this table presents the results of a bivariate regression. *p < .05; **p < .01; ***p < .001. The dependent variable is sensitive nuclear assistance coded from 0 (no assistance) to 1 (assistance). Robust standard errors are adjusted for clustering by dyad. For each independent variable, I report results for bivariate regressions with Rare Events Logistic Regression (ReL) and for bivariate regressions with Rare Events Logistic Regression after including spline corrections for temporal dependence (CS).

[1] *Superpower vote* and *Superpower vote (recipient)* are based on UNGA voting. Models that include these measures lose observations for dyads in which at least one of the countries was not a member of the UN. These models do not drop any observations in which sensitive nuclear assistance occurs with two exceptions. The China seat in the UN was held by Taiwan until 1971 and by China from 1971 to the present. Models using this variable do not capture French assistance to Taiwan in 1975 and Soviet assistance to China from 1958–1960. The other models using alternate measures of superpower dependence capture all cases of sensitive nuclear assistance.

Table 2.3. Correlates of sensitive nuclear assistance, 1951–2000: Multivariate results

INDEPENDENT VARIABLES		MODEL				
		1	2	3	4	5
Strategic	Power projection	−25.301*	−33.194***	−46.072***	−46.379***	−47.389***
		(11.307)	(9.895)	(13.099)	(14.246)	(9.728)
	Enemy	1.7429***	2.090***	1.986***	2.039***	1.497**
		(0.511)	(0.586)	(0.549)	(0.611)	(0.505)
	Superpower pact	−1.334***	−1.576**			
		(0.404)	(0.565)			
	Superpower vote			−1.314***	−1.261**	
				(0.323)	(0.420)	
	Nuclear weapon					1.101**
						(0.398)
Economic	Economic development	0.031		−0.034		−0.035
		(0.085)		(0.068)		(0.059)
	Growth	3.597		7.240*	9.701*	4.799*
		(3.345)		(2.896)	(3.899)	(2.447)
	Openness	−0.010		−0.094		−0.019
		(0.088)		(0.080)		(0.078)
	Trade dependence	44.420***	47.558***	50.160***	60.538***	53.275***
		(10.872)	(12.124)	(10.766)	(11.898)	(11.760)
Institutional	Regime type	−0.043		−0.059		−0.027
		(0.042)		(0.032)		(0.034)
	NPT	−1.053		−2.118*	−1.662*	−1.672*
		(0.647)		(0.910)	(0.778)	(0.847)
	NSG	2.292*	1.572*	3.597***	2.378***	3.163***
		(0.917)	(0.672)	(1.080)	(0.688)	(0.890)

(continued)

Table 2.3. Correlates of sensitive nuclear assistance, 1951–2000: Multivariate results (Continued)

INDEPENDENT VARIABLES		1	2	3	4	5
Strategic controls	Distance	19.818*	18.264*	20.852**	22.503**	18.779*
		(8.490)	(8.375)	(7.790)	(8.256)	(7.776)
	Distance squared	-1.259*	-1.156*	-1.303**	-1.394**	-1.174*
		(0.532)	(0.528)	(0.492)	(0.517)	(0.490)
Recipient demand	Disputes (recipient)	0.157		0.474***	0.444***	0.483***
		(0.134)		(0.118)	(0.125)	(0.115)
	Superpower pact (recipient)	0.457				
		(0.858)				
	Superpower vote (recipient)			-1.596*	-1.355	-2.322**
				(0.759)	(0.759)	(0.757)
	Economic development (recipient)	0.023		0.035		0.026
		(0.020)		(0.023)		(0.024)
	Liberalization (recipient)	0.008		-0.012		-0.011
		(0.022)		(0.017)		(0.017)
	Openness (recipient)	-0.014**	-0.011*	-0.011**	-0.011**	-0.012*
		(0.005)	(0.005)	(0.004)	(0.004)	(0.005)
	NPT (recipient)	-0.016		-0.945		-0.472
		(0.690)		(0.526)		(0.520)
	Constant	-81.552*	-75.545*	-85.352**	-93.746**	-78.487*
		(34.033)	(33.259)	(30.625)	(32.682)	(30.712)
N		81,952	81,952	78,143	78,143	78,920

Note: *p < .05; **p < .01; ***p < .001. The dependent variable is sensitive nuclear assistance coded from 0 (no assistance) to 1 (assistance). Robust standard errors are in parentheses and are adjusted for clustering by dyad. The model is estimated after including spline corrections for temporal dependence (Beck, Katz, and Tucker 1998).

The second hypothesis focuses on the existence of a shared rival as an incentive for nuclear supplier states to provide sensitive nuclear assistance. By providing nuclear assistance to a state with which they share a common enemy, nuclear suppliers can impose strategic costs on a rival state. As a reminder, we should expect a positive relationship between the existence of a shared enemy and sensitive nuclear assistance. Turning first to the bivariate models, we find support for this hypothesis. The relationship between the enemy variable and sensitive nuclear assistance is positive and statistically significant in every one of the bivariate models. The inclusion of control variables does not alter this relationship. In the multivariate models, we again see that the sign on the coefficient for the *Enemy* variable is positive and statistically significant. The analysis reveals a powerful link between the existence of a common enemy and the patterns of sensitive nuclear assistance.

Next, I examine the hypothesis that states that are vulnerable to superpower pressure will be less likely to provide sensitive nuclear assistance. In the bivariate analysis, we find the expected negative correlation between *Superpower pact* and *Superpower vote* and sensitive nuclear assistance. We also find the expected positive correlation between *Nuclear weapon* and sensitive nuclear assistance. *Superpower vote* is statistically significant in the bivariate models with the controls for temporal dependence, and *Superpower pact* is statistically significant in the simple bivariate model. *Nuclear weapon* is also statistically significant in both specifications. The rest of the relationships, however, are not statistically significant. Once I control for other relevant variables, however, a discernable relationship emerges. In each and every one of the multivariate models, the variable indicating the superpower dependence of the supplier is statistically significant and in the expected direction. States that are vulnerable to superpower pressure are less likely to provide sensitive nuclear assistance.

The findings of this analysis do not support rival explanations for why states transfer sensitive nuclear materials and technology. Some have argued that nuclear weapon states have an incentive to limit the size of the nuclear club. Thus, we may expect that nuclear weapon states will be less likely to provide sensitive nuclear assistance. This hypothesis is not supported by the evidence. Both the bivariate and multivariate analyses reveal a strong relationship between nuclear weapons possession and sensitive nuclear assistance—but in the opposite direction. As was discussed in the above paragraph on superpower dependence, nuclear weapon states are more, not less, likely to provide sensitive nuclear assistance, undermining arguments that preferences on nuclear weapons issues can be explained by nuclear weapons possession.

The set of economic variables does not appear to offer much explanatory power either. It is often assumed that states with lower levels of economic development or economic growth will be more likely to provide sensitive nuclear assistance because they will be more willing to take extreme measures to improve their economic circumstances. This expectation is not met with much empirical support. The variable measuring the level of economic development of the supplier is in the expected direction and is statistically significant in the simple bivariate model, but does not reach statistical significance in any of the other bivariate or multivariate models.[20] Moreover, the sign on *Economic development* is inconsistent, switching from negative to positive under different model specifications. The variable for economic growth is not statistically significant in any of the bivariate models. It is statistically significant in three of the four multivariate models in which it is included, but the sign on the coefficient is in the unexpected direction, suggesting that states experiencing slow economic growth are actually less likely to provide sensitive nuclear assistance. Neither is there a consistent, statistically significant relationship between openness to the international economy and sensitive nuclear assistance. *Openness* is negative and statistically significant in the simple bivariate model but is not statistically significant in any of the other models. The coefficient for trade dependence is in the expected direction and is statistically significant in all of the bivariate models and multivariate models, suggesting that states are more likely to provide sensitive nuclear assistance to states with which they have a close trade relationship.

Institutional hypotheses find only limited support. The NPT variable is negative and statistically significant in one of two bivariate models and in three of the four multivariate models in which it is included. Membership in the NPT does appear to serve as a constraint on the behavior of the nuclear suppliers. I do not find support, however, for the argument that the creation of the NPT led to the establishment of an international norm against nuclear proliferation strong enough to discourage sensitive nuclear assistance. To test this hypothesis, I ran a number of models (not shown) that included a dummy variable coded "1" for every observation after 1968 (the year in which the NPT was opened for signature). If this normative argument is correct, we would expect that states would be less likely to provide sensitive nuclear assistance after the establishment of the NPT in 1968. This variable was not statistically significant and in the expected direction in any of the models in which it was included, suggesting that the existence of the NPT does not impose a significant normative constraint on sensitive nuclear exports.

The findings of the NSG variable also fail to support the idea that membership in the institutions of the nuclear nonproliferation regime dissuade sensitive

---

20. Using GDP, instead of GDP per capita, produces similar results.

nuclear exports. The variable for NSG membership is not statistically significant in any of the bivariate models. In the multivariate models, the NSG variable is statistically significant and has a positive coefficient, suggesting that, contrary to expectation, membership in the NSG may actually increase the likelihood that a state will export sensitive nuclear materials and technology. This finding cannot be explained by superior supply capabilities of NSG members because the analysis excludes states that are incapable of providing sensitive nuclear assistance. Instead, this result may be explained by three factors. First, due to its status as a nuclear cartel, the NSG may have failed to acquire international legitimacy and has therefore failed to impose a meaningful normative constraint on its members. Second, an adverse selection effect in the early days of the NSG may have initially brought in the states that were most likely to export sensitive technologies. These states joined the NSG but continued to provide sensitive nuclear assistance. For example, the United States convinced a reluctant France to join the regime in 1975, but France continued the construction of a pilot-scale plutonium reprocessing plant in Pakistan until 1982. Third, the guidelines of the NSG allow states to export sensitive nuclear materials and technology as long as the exports are placed under international safeguards. NSG states may be more likely to provide sensitive nuclear assistance, but they may also be more likely to place their exports under safeguards. For example, Germany joined the NSG in 1975 and required safeguards on its sensitive nuclear exports to Brazil from 1978 to 1994.

Next, I summarize the results of the other control variables. The variables for distance and distance-squared are statistically significant in each of the multivariate models, demonstrating a non-monotonic relationship between distance and sensitive nuclear assistance.[21] Turning to the demand-side variables, we see that states in a competitive security environment are more likely to receive sensitive nuclear assistance. The coefficient on the *Disputes* variable is positive and statistically significant in three of the four models in which it is included. We also find that states that are dependent on a superpower patron are less likely to receive sensitive nuclear assistance. The sign on *Superpower vote (recipient)* is negative and statistically significant in two of the three multivariate models in which it is included. There is also support for the argument that states that are open to the international economy have less demand for nuclear weapons and are thus less likely to receive sensitive nuclear assistance. The coefficient on *Openness (recipient)* is negative and statistically significant in each of the

---

21. I also tried the unlogged variables for distance and distance squared. The findings did not change.

Table 2.4. Substantive effects of variables on the likelihood of sensitive nuclear assistance, 1951–2000

VARIABLE	RELATIVE RISKS	95% CONFIDENCE INTERVALS
Power projection	2.894	1.630 to 5.457
Enemy	7.200	2.248 to 22.554
Superpower pact	4.791	1.526 to 15.573
Superpower vote	2.968	1.753 to 5.133
Trade dependence	1.360	1.195 to 1.547
NPT	8.559	1.620 to 47.546

Note: The probabilities are calculated using the ReLogit estimates in model 3 of table 2.3. Since the *Superpower pact* variable is not included in model 3, the probabilities for this variable are calculated using the ReLogit estimates in model 2 of table 2.3. The calculations were performed to allow the presentation of the relative risks as positive numbers. The entries for the *Power projection* and *Superpower vote* variables represent the effect of a change in the independent variable from one standard deviation above the mean to one standard deviation below the mean. The entries for the *Trade dependence* and *Disputes* variables represent the effect of a change from one standard deviation below the mean to one standard deviation above the mean. The entries for the *Enemy* and *NPT* variables represent the effect of a change from 0 to 1 on the dichotomous variable. The entries of the *Superpower pact* variable represent the effect of a change from 1 to 0 on the dichotomous variable. All other variables are held at their mean.

multivariate models. The other control variables are not statistically significant in any of the models.[22]

We have seen that the variables identified by the strategic theory of nuclear proliferation are statistically significant, but how important are they substantively? Table 2.4 presents the substantive effect of these variables on sensitive nuclear assistance.[23] The entries represent the relative risks of sensitive nuclear assistance for a given change in the independent variable when all other variables are held at their mean.[24] Turning first to *Power projection*, the table reveals that a state that lacks the ability to project power over a potential recipient is nearly three times more likely to provide sensitive nuclear assistance to that state than a comparable state that has the ability to project power over the potential recipient.

---

22. I also tried models that included variables for the regime type of the recipient, joint democracy of the supplier and the recipient, population size of the supplier and recipient, the number of nuclear weapon states in the international system, and a dummy for the Cold War period. None of these variables achieved statistical significance or altered the core findings.

23. Tests for substantive effects are performed using the results from table 2.3, models 2 and 3. Probabilities are calculated using the approximate Bayesian method for predicting probabilities in rare events as recommended by King and Zeng, "Logistic Regression in Rare Events Data."

24. Table 2.4 also reports the 95 percent confidence bounds using the procedure developed by King, Tomz, and Wittenberg. Gary King, Michael Tomz, and Jason Wittenberg, "Making the Most of Statistical Analyses," *American Journal of Political Science,* Vol. 44, No. 2 (April 2000), pp. 341–355.

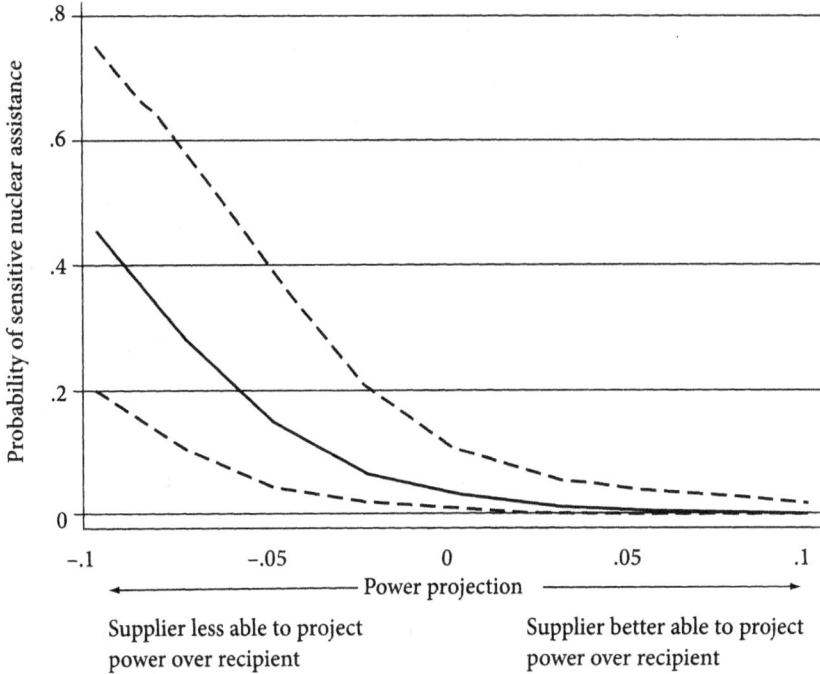

FIGURE 2.1. Effect of power projection on sensitive nuclear assistance, 1951–2000.

Note: The probabilities are calculated using the ReLogit estimates in model 2 of table 2.3. *Enemy* is set to one, *Superpower pact* is set to zero, and all other variables are held at their mean. *Relative power* is measured as the power of the supplier, discounted by the distance between the supplier and the recipient, minus the power of the recipient. The unit of measurement is the proportion of overall power in the international system. A *Relative power* score of −0.1 indicates that, after discounting by distance, the supplier possesses 10 percent less of the total power in the international system than the recipient. Negative values of *Relative power* suggest that the supplier lacks the ability to project power over the recipient. Dyad-years in this category include France-Japan 1973 (−0.059) and Pakistan–North Korea, 1999 (−0.011). A *Relative power* score of 0.1 indicates that, after discounting by distance, the supplier possesses 10 percent more of the overall power in the international system than the potential recipient. Positive values of *Relative power* suggest that the supplier possesses the ability to project power over the recipient. Dyad-years in this category include France-Belgium 1960 (0.015) and the United States–North Korea 1951 (0.039). Dashed lines represent the 95 percent confidence interval.

This effect of *Power projection* on the probability of sensitive nuclear assistance is illustrated in figure 2.1. In this figure, *Enemy* is set to one, *Superpower pact* is set to zero, and all other variables are held at their mean. The dashed lines represent the 95 percent confidence interval. The figure demonstrates a clear negative relationship between power projection and the probability of sensitive nuclear assistance. When the capable nuclear supplier is unable to project power over the potential recipient (represented in the graph by negative values of *Power projection*), there is a substantial risk that the supplier state will export sensitive nuclear materials and technology. As the relative power distribution shifts in favor of the potential supplier, however, the risk that a state will export sensitive nuclear materials and technology declines. The probability of sensitive nuclear assistance decreases nearly to zero as the capable nuclear supplier gains the ability to project power over the potential nuclear recipient (represented in the graph by positive values of *Power projection*).

Returning to the substantive effect of the other variables reported in table 2.4, we see that states are over seven times more likely to provide sensitive nuclear assistance to a state with which they share a common rival than to a comparable state with which the supplier does not share a common rival. And a state that is not dependent on a superpower patron as measured by the superpower pact variable is nearly five times more likely to provide sensitive nuclear assistance than a similar state that is dependent on a superpower patron. Furthermore, a state that is not dependent on a superpower patron, as measured by UNGA voting data, is about three times more likely to provide sensitive nuclear assistance than a similar state that is dependent on a superpower.

The substantive importance of these variables is also evident when they are taken together (not shown). Sensitive nuclear assistance, a rare event, is extremely unlikely to occur under typical circumstances. Indeed, the baseline probability that a nuclear transaction will occur in any given dyad year when all of the explanatory variables are held at their mean is 0.2 percent.[25] When the three strategic variables are set to the worst-case scenario (that is, *Power projection* and *Superpower pact* set to their minimums and *Enemy* set to its maximum), however, and all other variables are held at their mean, the probability of sensitive nuclear assistance increases to 62 percent.[26] Taken together, this analysis indicates that strategic factors have not just a statistically significant effect but also a substantively significant effect on decisions to provide sensitive nuclear assistance.

Turning now to the substantive effect of the control variables, table 2.4 shows that NPT membership has a substantive effect on the patterns of sensitive nuclear assistance. Non-NPT members are over eight times more likely to provide sensitive nuclear assistance than are NPT member states. By contrast, trade dependence has a small substantive effect on states' propensity to export sensitive nuclear technologies. States that are heavily dependent on trade with a particular trading partner are only about 36 percent more likely to provide sensitive nuclear assistance to that state than is a similar state that has no trade relationship whatsoever with that state. Thirty-six percent may not be a significant difference

---

25. The 95 percent confidence interval is 0.001 to 0.005. The probabilities reported here are calculated using the ReLogit estimates in table 2.1, Model 2.

26. The 95 percent confidence interval is 0.125 to 0.945.

when one considers that a state with very high levels of trade with a particular state should be much more likely (indeed, more than 36 percent more likely) to trade any product with that state than with a similar state with which the supplier has absolutely no trade relationship.

Conclusion

In this chapter, I presented the results of a quantitative analysis on the patterns of sensitive nuclear assistance. The results are clear. Strategic factors are the most important determinants of sensitive nuclear assistance. The inability to project power over the potential recipient, the presence of a common enemy, and superpower independence—all increase the risk of sensitive nuclear transfers.

This chapter was unable to find evidence that economic concerns motivate the provision of sensitive nuclear assistance. There is not a discernable negative relationship between either economic development or economic growth and sensitive nuclear assistance. Economic hardship does not appear to increase the risk that a state will export sensitive nuclear materials or technology. Neither do other economic factors exert a statistically and substantively significant effect on the trade in sensitive nuclear wares. These results cast serious doubts on the claims that sensitive nuclear exports are driven by economic factors.

The alternate hypothesis that the nuclear nonproliferation regime will affect patterns of sensitive nuclear transfers found some support. There was evidence to suggest that states that are members of the NPT are less likely to provide sensitive nuclear assistance. This finding reassures those who believe that the nuclear nonproliferation regime has been a critical factor halting the proliferation of nuclear weapons. Unfortunately for supporters of the nuclear nonproliferation regime, the analysis did not find that all of the institutions of the regime exert a similar restraining effect. We found that states that are members of the NSG are actually more likely to transfer sensitive nuclear technologies than are capable nuclear suppliers that are not members of the NSG. Furthermore, there is no evidence to suggest that the mere existence of the NPT imposes a normative constraint on sensitive nuclear exports. States are not less likely to provide sensitive nuclear assistance in the post-NPT era.

The statistical analysis performed in this chapter did not provide support for the idea that nuclear weapon states will be less likely to provide sensitive nuclear assistance. Rather, consistent with the expectations of the strategic theory of nuclear proliferation, I found that states that possess nuclear weapons are more likely to export sensitive nuclear materials and technologies. It is also possible that nuclear weapon states are more likely to provide sensitive nuclear assistance because they can better provide assistance related to weaponization or because they believe they can better deter nuclear attacks.

Having explained the factors that account for the aggregate pattern of sensitive nuclear assistance, I now explore the mechanisms that link strategic conditions to sensitive nuclear transfers. To this end, the next chapter presents an in-depth analysis of an important case of sensitive nuclear assistance.

3

ISRAEL'S NUCLEAR PROGRAM
French Assistance and U.S. Resistance

From 1959 to 1965, France provided Israel with sensitive nuclear assistance, helping Israel acquire nuclear weapons. During the same time period, the United States refused Israeli requests for sensitive nuclear assistance and actively intervened in an attempt to prevent French-Israeli nuclear cooperation. What explains the different approaches that France and the United States took to the Israeli nuclear program? Why did France help Israel acquire nuclear weapons? Why was the United States determined to prevent Israel from becoming a nuclear power?

In this chapter, I argue that the answers to these questions can be found in the strategic theory of nuclear proliferation. The United States enjoyed a force-projection capability over Israel and feared that nuclear proliferation in the Middle East would undermine America's strategic position in the region. For this reason, the United States refused to provide Israel with sensitive nuclear assistance and acted to prevent French-Israeli nuclear cooperation. The French strategic calculation was quite different. France's inability to project military power over Israel meant that France would not incur significant strategic costs in the event that Israel acquired nuclear weapons. Moreover, French policymakers calculated that a nuclear Israel would improve France's strategic position in the region by constraining a key enemy. French officials hoped that by strengthening Israel they could impose strategic costs on Egypt, a country that was aiding the forces fighting France in the Algerian civil war. France considered America's opposition to nuclear proliferation to Israel but was able to ignore Washington's protests and continue with the sensitive nuclear transaction because France did not depend on the United States to meet its core security needs. Alternative explanations based on economics, institutions, or nuclear weapons possession do not find support in this case.

Background and Overview
The Origin of Israel's Nuclear Program

Israel's desire for nuclear weapons can be traced to the country's founding.[1] Upon independence in 1948, Israeli officials were faced with an extremely threatening security environment. In the 1948 Arab-Israeli War, Israel was invaded by a coalition of Arab states.[2] Israel survived the war, but David Ben-Gurion, Israel's first prime minister, and other key Israeli officials anticipated future conflicts. Israeli officials believed that deterring a second Arab invasion with conventional military forces would be difficult, if not impossible, given the Arab world's vast superiority in population and resources. While building up conventional military forces, Israel simultaneously began to pursue the protection that could be provided by a nuclear weapons arsenal. David Ben-Gurion sought to develop the nuclear science that eventually could contribute to a nuclear deterrent; as part of this program, he established the Israeli Atomic Energy Commission (IAEC) in 1952.

The commission was headed by Ernst David Bergmann, a leading Israeli chemist.[3] Bergmann represented a minority among Israel officials who believed that Israel could develop nuclear weapons without external assistance. Other key players in the Israeli defense establishment, however, concluded that Israel's nuclear development would require foreign help.[4] Shimon Peres (at the time a rising star in the Ministry of Defense) calculated that Israel's security situation demanded nuclear weapons immediately and that Israel could not afford to wait to develop an autonomous program. Led by Peres, Israel began an international search for a nuclear supplier.

---

1. This section draws heavily on a number of important studies on Israel's nuclear program that include information about the French role in Israel's nuclear development. These studies include: Cohen, *Israel and the Bomb;* Hersh, *Samson Option;* Pierre Péan, *Les Deux Bombes* (Paris: Fayard, 1981); and Benjamin Pinkus, "Atomic Power to Israel's Rescue," *Israel Studies,* Vol. 7, No. 1 (2002), p. 119.

2. On the history of the 1948 Arab-Israeli War, see, for example, Uri Milstein, *History of the War of Independence,* vols. 1–4 (New York: University Press of America, 1996).

3. On the establishment of the IAEC, see, for example, Shimon Peres, *From These Men* (New York: Wyndham, 1979).

4. Shimon Peres, *Battling for Peace* (London: Weidenfeld and Nicolson, 1995), pp. 133–135.

Atoms for Peace and the U.S. Approach

Israel identified the United States as a potential nuclear supplier in December 1953 when President Dwight D. Eisenhower delivered the Atoms for Peace speech in the United Nations General Assembly.[5] Under the Atoms for Peace program, the United States promised to provide basic nuclear science and technology to developing countries. Atoms for Peace was partly an effort to harness nuclear energy to aid economic development in the Third World, partly an attempt to expand U.S. influence in the Cold War competition against the Soviet Union, and partly an early U.S. nuclear nonproliferation strategy. By satisfying states' demand for nuclear technology for energy and other peaceful purposes, the Eisenhower administration hoped that it could dissuade states from developing military-oriented nuclear programs.

Contrary to the nonproliferation aims embodied in the Atoms for Peace program, however, Israeli officials viewed Eisenhower's offer as an opportunity to jump-start their fledgling nuclear weapons program. Israel was the second country (after Turkey) to sign up to Atoms for Peace. The move paved the way for the 1955 U.S.-Israel nuclear cooperation agreement under which the United States promised to provide Israel with a nuclear research reactor.[6] Under the terms of the agreement, a system of safeguards was established, and the nuclear assistance was designated for peaceful purposes.

The research reactor helped Israel take its first steps toward the development of a nuclear program. Yet, Israeli scientists understood that a small research reactor could not contribute directly to nuclear weapons. To develop a nuclear arsenal, Israel would require much more advanced nuclear technologies. Shortly thereafter, Israel sought to upgrade its nuclear cooperation agreement with Washington. At the Geneva Conference on the Peaceful Use of Atomic Energy in August 1955, Bergmann approached the U.S. delegation and asked for more sophisticated nuclear technology.[7] Bergmann asked the Americans for "something like a real reactor." Bergmann told the Americans that this was necessary so that Israel could produce "new elements, such as plutonium."[8]

---

5. For a book-length treatment of the Atoms for Peace Program, see, for example, Richard G. Hewlett and Jack M. Holl, *Atoms for Peace and War, 1953–1961* (Berkeley: University of California Press, 1989).

6. *Atomic Energy* (Washington, D.C.: Department of State, publication 5963), as cited in Cohen, *Israel and the Bomb*, p. 45.

7. "The Geneva Conference on the Peaceful Use of Atomic Energy (8–20 August 1955)," signed by A. D. Bergmann, ISA, FMRG, 2407/2 (the top-secret version), as cited in Cohen, *Israel and the Bomb*, p. 45.

8. Ibid.

The United States was unwilling, however, to contribute further to Israel's nuclear development. Admiral Lewis Strauss, the head of the U.S. delegation and chair of the U.S. Atomic Energy Commission (AEC), flatly rejected Bergmann's request. The United States was opposed to Israel producing elements that could potentially be used in a nuclear weapons program. Strauss warned Bergmann that the United States would not do "anything that provided you even the slightest quantities of plutonium." Strauss continued to express America's intention to prevent civilian nuclear cooperation with Israel from feeding into a military program. Strauss threatened Bergmann that, when it came to U.S.-Israeli nuclear cooperation, "there would be control."[9] Just as it was becoming apparent that the United States would be unwilling to provide sensitive nuclear assistance to Israel, an alternate nuclear supplier emerged.

Nasser and the French Approach

In the late 1950s and early 1960s, France's principal foreign policy challenge was to defeat the insurgency in French Algeria and to prevent Algeria from gaining independence.[10] French officials had reason to believe that President Gamal Abdel Nasser of Egypt was the principal external supporter of the main rebel organization in Algeria, the Front de Libération Nationale (FLN). On October 20, 1956, France captured the Egyptian vessel *Taulus* transporting arms to the Algerian rebels. French officials concluded that severing Egyptian support to the rebels was a necessary condition for France to achieve victory in Algeria. In fact, French officials believed that countering Nasser was France's foremost strategic objective, more important than achieving tactical victories in Algeria itself. French Foreign Minister Christian Pineau even declared, "France considers it more important to defeat Colonel Nasser's enterprise than to win ten battles in Algeria."[11] French officials set out to constrain Nasser's growing power and influence in the Middle East and, according to the fantasies of some French officials, overthrow Nasser's regime.

French officials calculated that the best way to counter Nasser's influence in the region was to enhance the military power of his principal military rival,

---

9. Ibid.
10. On the Algerian War, see, for example, Alistair Horne, *Savage War of Peace* (New York: NYRB Classics, 2006); Martin S. Alexander, Martin Evans, and J. F. V. Keigler, eds., *The Algerian War and the French Army, 1954–1962* (New York: Palgrave Macmillan, 2002); and Martin S. Alexander, *France and the Algerian War, 1954–1962* (New York: Routledge, 2002).
11. Christian Pineau, as quoted in William Roger Louis and Roger Owen, eds., *Suez 1956* (Oxford: Clarendon, 1989), p. 137.

Israel.[12] According to historian Avner Cohen, "A militarily strong Israel, capable of threatening Nasser, was now in France's interest."[13] A strengthened Israel would force Nasser to limit his regional ambitions and shift his strategic attention toward Israel and away from Algeria.

France's strategy to bolster Israeli power vis-à-vis Egypt was threatened in late September 1955, however, when Nasser announced a conventional military arms deal between Egypt and the government of Czechoslovakia. The sale, approved by Egypt's external patron, the Soviet Union, would transfer artillery, armor, and aircraft from the Soviet bloc to Egypt.[14] In less than a year, Egypt's conventional military armaments would more than double in both quantity and quality. The Czech-Egyptian arms sale would alter the military balance of power in the region and potentially threaten France's regional strategy.

In response, France became, within months of the Czech-Egyptian arms sale, Israel's largest provider of conventional military hardware.[15] Beginning in 1955, France provided Israel with *Mystere* jet fighters, AMX tanks, 155-millimeter guns, and other military equipment. French and Israeli strategic cooperation further intensified on July 26, 1956, when President Nasser announced the nationalization of the Suez Canal.[16] France saw an opportunity to take military action against Nasser, but the operation would require Israeli participation. On the very next day, French Defense Minister Maurice Bourgès-Maunoury summoned Peres to a meeting and proposed a tripartite military operation to retake the Suez Canal.[17] According to the plan, Israel would invade the Sinai Peninsula. After Israel had secured Sinai, France and Great Britain would call for a ceasefire and the establishment of a demilitarized zone on either side of the Suez Canal. When Egypt refused (which the co-conspirators believed was inevitable), French and British forces would invade and occupy the canal themselves under the auspices of maintaining a ceasefire between Egypt and Israel. France and Britain could

---

12. On the French-Israeli strategic relationship, see, for example, Sylvia Crosbie, *A Tacit Alliance* (Princeton, NJ: Princeton University Press, 1974); Shimon Peres, *David's Sling* (London: Weidenfeld and Nicolson, 1970); and Michael Brecher, *Decisions in Israel's Foreign Policy* (New Haven: Yale University Press, 1975).

13. Cohen, *Israel and the Bomb*, p. 49.

14. On Soviet-Egyptian relations in this period, see, for example, Karen Dawisha, *Soviet Foreign Policy toward Egypt* (New York: St. Martin's, 1979).

15. On Israel's search for arms suppliers in this period and on French arms sales to Israel, see Peres, *Battling for Peace*.

16. For a full discussion of the Suez Crisis, see, for example, Selwyn Ilan Troen and Moshe Shemesh, eds., *The Suez-Sinai Crisis 1956* (New York: Columbia University Press, 1990), A. J. Barker, *Suez* (London: Faber and Faber, 1964); Anthony Gorst and Lewis Johnman, *The Suez Crisis* (London: Routledge, 1997); Louis and Owen, eds., *Suez 1956;* Terrence Robertson, *Crisis* (Toronto: McClelland and Stewart, 1964); and Brecher, *Decisions in Israel's Foreign Policy.*

17. Peres, *Battling for Peace*, p. 53.

then wrest control of the Suez Canal back from Egypt and, perhaps, precipitate the collapse of Nasser's regime.

It was in this context that France took its first steps toward aiding Israel's nuclear ambitions. On September 13, 1956, Bergmann approached France with exactly the same request that he had put to the Americans just one year before. According to Bertrand Goldschmidt, head of the Commissariat l'Énergie Atomique (CEA), Bergmann and Peres "asked for [French] help in reaching nuclear capability, that is, to supply enough basic elements to produce plutonium, obtain a reactor, and, if necessary, to generate electricity."[18] Unlike the United States, however, France agreed, pledging to provide Israel with a second nuclear reactor. The proposed reactor, modeled on France's 18-megawatt EL-3 reactor at Saclay, was relatively small, though it would have allowed Israel to produce plutonium. The exchange was not a specific quid pro quo for Israeli participation in the Suez plan but was part and parcel of a growing French strategy to strengthen Israel as a means of constraining Egypt.[19]

On October 29, 1956, the Suez Crisis began as Israel invaded the Sinai Peninsula. On November 6, British and French forces followed Israel into combat with the planned amphibious invasion of Egypt. The operation soon came to a screeching halt, however, in the face of an unexpected intervention from the Soviet Union. The USSR demanded that the hostilities cease and backed up its demand with the threat of military action. The Soviet Union threatened all three states with nuclear retaliation but singled out Israel with its gravest warning. The Soviet Union accused Israel of "irresponsibly playing with the fate of its own people...which puts in jeopardy the very existence of Israel as a state."[20] In a separate letter, Soviet Prime Minister Nikolai Bulganin reminded Ben-Gurion that the USSR was capable of striking Israel with nuclear-armed missiles. As if the threat were not already clear enough, the Soviets also expressed the possibility that Soviet ground troops, or "volunteers," would be sent to the region.[21]

In the face of the Soviet threats, Israel sought international protection. The United States made it clear that it would be unwilling to defend Israel in the event of a Soviet intervention.[22] Instead, President Eisenhower joined the Soviet Union in outspoken opposition to Israel's unprovoked military adventure. The

---

18. Pinkus, "Atomic Power to Israel's Rescue," p. 119.
19. Yossi Melman, interviewed by Shimon Peres, in "Royal Gift" (in Hebrew), *Ha'artz*, October 11, 1991, as cited in Cohen, *Israel and the Bomb*, p. 363.
20. "Soviet Protests Canal Blockade," *New York Times*, November 5, 1956.
21. Robertson, *Crisis*, p. 252.
22. There is a scholarly debate about which factor, the Soviet threats or U.S. opposition, was more important in shaping Israel's decision to withdrawal from Suez. Advocates on both sides, however, admit that both factors were important. See, for example, Brecher, *Decisions in Israel's Foreign Policy*, pp. 225–317.

United States also demanded an immediate ceasefire and the withdrawal of the French, British, and Israeli troops from Egyptian territory.

Unable to rely on the United States, Israel again turned to France. On November 8, less than two days after the Soviet threats, Ben-Gurion dispatched Simon Peres and Golda Meir to Paris to "find out what the French stand would be in the event of specific Soviet intervention," according to Peres.[23]

The unexpected outcome of the Suez Crisis threatened France's strategic goals. France had not succeeded in overthrowing Nasser's regime; French and British troops were forced to withdraw from the region; and Nasser, though militarily weakened, was still in power and was politically strengthened by his apparent victory in the Suez Crisis. Further, the Soviet Union's strong military and political support for Egypt and antagonism to Israel threatened to weaken France's key strategic partner in the region and the cornerstone of France's Algeria strategy.

In Paris, the Israeli delegation met with Bourgès-Maunoury and French Foreign Minister Christian Pineau. The French officials urged Israel to comply with the Soviet demands.[24] French officials confessed to Meir and Peres that France lacked the capability to challenge the Soviet Union militarily in the Middle East but explained that they maintained a strong strategic interest in guaranteeing Israel's security. According to Peres, "Pineau, more dour than ever, said that France stood shoulder-to-shoulder with the Jewish state…but they had nothing that could shoot down Soviet missiles."[25]

When Peres requested that France help Israel acquire nuclear weapons, the French were willing to help. On November 8, France promised to help Israel acquire nuclear weapons.[26] France agreed to provide Israel with a 40-megawatt nuclear reactor capable of generating ten to fifteen kilograms of plutonium a year.[27] This reactor's output was more than double the 18-megawatt reactor that France had agreed to provide in September 1956. Even more significantly, the nuclear package included an underground plutonium reprocessing plant. The reprocessing plant would give Israel the capability to reprocess the spent fuel from the nuclear reactor and separate weapons-grade plutonium that could be used in the core of an atomic bomb. Construction of the reactor began in 1958, and construction of the sensitive reprocessing facility was initiated in the following

---

23. Peres, *David's Sling*, p. 211.
24. Michael Bar-Zohar, *Ben Gurion* (Tel Aviv: Smora Bitan, 1987), pp. 1271–1273; and Peres, *Battling for Peace*, p. 131.
25. Shimon Peres, "The Road to Sèvres," in Troen and Shemesh, eds., *Suez-Sinai Crisis 1956*, p. 131.
26. See, for example, Péan, *Les Deux Bombes*, p. 83; and Reed and Stillman, *Nuclear Express*, pp. 68–83.
27. Cohen, *Israel and the Bomb*, pp. 58–60, 73; Péan, *Les Deux Bombes*, pp. 96, 126–128; and "France Admits It Gave Israel A-Bomb," *Sunday Times* (London), October 12, 1986.

year. During this time, it is believed that France also assisted Israel with nuclear weapon designs and allowed Israeli observers at French nuclear tests.

France and Israel conducted the negotiations and subsequent nuclear transactions with the utmost secrecy. The French government was eager to keep the unprecedented level of international nuclear cooperation away from the criticism of an international audience.[28] To ensure the secrecy of the transaction, a special front organization was created for the transfer of funds between the two countries and, to provide the French government plausible deniability, the contract for the plutonium reprocessing plant, the most sensitive aspect of the transaction, was signed directly between Israel and Saint-Gobain Techniques Nouvelles (SGN), a private French contractor.

U.S. Suspicions and Pressure

The United States did not initially suspect Israel to be a nuclear proliferation threat. In fact, in a 1957 NIE, the U.S. intelligence community concluded that Israel was unlikely to acquire nuclear weapons in the short term because Israel lacked the indigenous technical capability to produce nuclear weapons. The estimate concluded that Israel "would require major foreign assistance to produce even the first nuclear weapon within the next 10 years."[29] By the late 1950s and early 1960s, however, the United States began to acquire intelligence that hinted that Israel was getting such help and expanding its nuclear research program.[30] Most notably, in early 1958, a U-2 spy plane obtained images of a large facility under construction in the Negev Desert. While the signatures of the site were inconclusive, CIA analysts speculated that the site could be nuclear-related.[31]

Following the U-2 photographs, the U.S. intelligence community undertook a program to gather more systematic evidence on the construction site. For example, the CIA station in Tel Aviv provided picnic baskets, bottles of wine, and special cameras to U.S. diplomatic personnel willing to picnic near the Dimona site, take a few pictures, and collect earth samples.[32]

As the intelligence indicating that Israel was secretly expanding its nuclear program, possibly with French assistance, continued to accumulate, the United

---

28. Cohen, *Israel and the Bomb*, pp. 59, 60; and Gabriel Sheffer, *Moshe Sharret* (Oxford: Oxford University Press, 1996), pp. 808–859.
29. Director of Central Intelligence, NIE 100-6-57, *Nuclear Weapons Production in Fourth Countries,* June 18, 1957, p. 4, as cited in Richelson, *Spying on the Bomb,* p. 247.
30. For a complete account of U.S. efforts to gather intelligence on Israel's nuclear program see Richelson, *Spying on the Bomb.*
31. Ibid, pp. 248–250.
32. Ibid.

States began to confront French and Israeli officials with their suspicions. At first, the suspects flatly denied any sensitive nuclear cooperation. When U.S. officials challenged Bergmann about the site, he admitted that Israel and France had reached a nuclear cooperation agreement but gave assurances that the deal was limited to the exchange of information on uranium chemistry and the production of heavy water.[33]

The United States also demanded information from the French government. In November 1960, for example, the AEC representative in Paris confronted a French CEA official with charges that France was helping Israel construct a nuclear power plant. The CEA official "stated flatly" that neither France nor Israel was constructing a nuclear reactor in Israel and repeated Bergmann's story that French-Israeli nuclear cooperation was limited to uranium mining and heavy water production.[34]

While French officials denied any wrongdoing in public, they seriously considered the possible negative repercussions of their actions in private. Newly elected president Charles de Gaulle approved of France's support of Israel's nuclear program but is reported to have worried that "if France was the only country to help Israel, while neither the United States, Britain or the Soviet Union has helped anyone else [get the bomb], she would put herself in an impossible international situation."[35] De Gaulle's fears of international condemnation were not enough, however, to cause him to cut off France's nuclear assistance to Israel. Rather, de Gaulle merely took steps to cover up the official role of the French government in assisting Israel's nuclear program. In November 1960, de Gualle ordered a halt to the CEA's direct nuclear cooperation with Israel but encouraged French firms to continue the construction of the nuclear reactor (which at this point was nearly complete) and the plutonium reprocessing facility.

According to a version of events popularized by investigative journalist Seymour Hersh, de Gaulle ordered a halt to all French assistance to Israel in 1960, but the French bureaucracies continued the sensitive nuclear cooperation with Tel Aviv behind de Gaulle's back without the government's knowledge or official authorization. While Hersh's account makes for an arresting story, the preponderance of evidence in other accounts demonstrates that de Gaulle made a calculated decision to cut off official cooperation to provide the French government

---

33. "Post-Mortem on SNIE 100-8-60: Implications of the acquisition by Israel of a Nuclear Weapons Capability," Draft, January 31, 1961, Department of State Lot files, Lot No57D688, USNA, as cited in Cohen, *Israel and the Bomb*, p. 82.
34. Telegram (2612), Amory Houghton to Secretary of State, November 26, 1960, Nonproliferation Collection, as cited in Cohen, *Israel and the Bomb*, p. 86.
35. Charles de Gaulle, as cited in Hersh, *Samson Option*, p. 69.

plausible deniability while simultaneously utilizing private French firms to honor the nuclear cooperation agreement with Israel.[36]

Inconsistencies in the official French and Israeli stories eventually revealed the full scope of French-Israeli nuclear cooperation. When the United States continued to confront Israel with intelligence that suggested that French-Israeli cooperation included a nuclear reactor, various levels of the Israeli government responded with drastically different cover stories. Some claimed that Dimona was a metallurgical research facility while others maintained that it was a textile plant.[37] On December 18, 1960, AEC chairman John McCone appeared on the NBC television show *Meet the Press* and announced that the United States had reason to believe that Israel was secretly constructing a nuclear reactor with French assistance.[38] The very next day, France's Ministry of Foreign Affairs and Israel's Embassy in Washington D.C. released official statements that, for the first time, acknowledged French-Israeli cooperation on the construction of a nuclear reactor. The statements also pledged, however, that the reactor was for civilian, not military, use and "that all necessary provisions have been taken by France to assure that the French aid to Israel in the nuclear field would be used only for peaceful purposes."[39] At this point, the world was not yet aware of the plutonium reprocessing facility being constructed beneath the Dimona site, and neither the French nor the Israel statements mentioned the construction of the underground facility.

Still, the United States doubted the peaceful intentions of the French-Israeli nuclear deal. Secretary of State Dean Rusk notified President-Elect John F. Kennedy that it was "the intention of our intelligence agencies to maintain a continuing watch on Israel as on other countries to assure that nuclear weapons capabilities are not being proliferated."[40] Intelligence analysts soon began to wonder whether the Dimona site might also contain a plutonium reprocessing facility.[41] After all, the Dimona reactor was based on the French reactor at

---

36. For example, see Peres, *Battling for Peace*, p. 142; Cohen, *Israel and the Bomb*, p. 75; and Reed and Stillman, *Nuclear Express*, pp. 80, 117.

37. "Post-Mortem on SNIE 100-8-60"; and "Memorandum for the President: Dimona Reactor in Israel," an attachment: "History of the United States Interest in Israel's Atomic Activities," March 30, 1963, RG 59, Central Foreign Policy Decimal Files, 250/03/27/04, Box 1297, 611.84a45/3-3061, USNA, as cited in Cohen, *Israel and the Bomb*, p. 85.

38. John W. Finney, "U.S. Hears Israel Moves Toward A-Bomb Potential," *New York Times*, December 19, 1960, p. 1.

39. "Ben Gurion Explains Nuclear Project," *New York Times*, December 19, 1960, p. 8; and State Department Compilation of "Political Statements Concerning the Israeli Reactor," January 17, 1961, 3207, as cited in Cohen, *Israel and the Bomb*, p. 92.

40. Dean Rusk, as cited in Richelson, *Spying on the Bomb*, p. 254.

41. Cohen, *Israel and the Bomb*, p. 58; Péan, *Les Deux Bombes*, p. 96; Richelson, *Spying on the Bomb*, p. 249; and "Memorandum of Discussion with Mr. Pratt and Staff," attached to "Notes on Visit

Marcoule, which itself was placed next to a reprocessing plant. U.S. suspicions about the existence of a reprocessing facility were only heightened when Israel officials publicly mentioned Israel's desire to acquire, at some future point in time, a plutonium reprocessing laboratory.

The United States once again pressured both countries for a confession, this time related to a reprocessing facility. It is reported that at a Washington dinner party in 1962, John McCone, who by this time had become director of the CIA, pointedly queried Charles Lucet, a senior French Foreign Ministry official, about the suspected reprocessing facility: "So Mr. Lucet, your country is building a reprocessing plant for the Israelis?" Lucet denied the accusation, repeating France's official position: "No, we are building a reactor."[42]

The United States felt that more verification was needed to confirm that French-Israel nuclear cooperation was limited to a nuclear reactor. The United States demanded inspections to verify that the reactor was intended for peaceful purposes and that the Dimona site did not contain a reprocessing facility. In one of his first official acts as president, Kennedy wrote to Ben-Gurion requesting that American scientists be allowed to inspect the Dimona nuclear facility.[43] Kennedy sought a formal agreement that would allow for semiannual inspections. When Ben-Gurion put up initial resistance, Kennedy followed up with subsequent correspondence that went so far as to threaten to put U.S.-Israeli relations on the line over the nuclear issue. In three separate letters to Ben-Gurion, Kennedy used the same language, stressing that "this government's commitment to and support of Israel could be seriously jeopardized if it should be thought that we were unable to obtain reliable information on a subject as vital to peace as the question of Israel's effort in the nuclear field."[44]

The Israelis had obvious reasons to resist Kennedy's proposal. Intrusive international inspections would expose the extent of French-Israeli nuclear cooperation and jeopardize Israel's nuclear weapons program. Over the next seven years, the Israeli response to repeated U.S. requests for inspections followed a pattern.[45] Ben-Gurion, and his successor Levi Eshkol, would not explicitly reject U.S. requests for inspections. Rather, they would reiterate the peaceful intentions of the nuclear program and continually delay inspections, citing the need for secrecy and a host of domestic political concerns.

---

to Israel, U. M. Staebler–J. W. Croach Jr.," DRAFT 5/23/61, as cited in Cohen, *Israel and the Bomb*, p. 106.

    42. Hersh, *Samson Option*, pp. 118, 119.
    43. Ibid., pp. 100–101.
    44. See Warren Bass, *Support Any Friend* (Oxford: Oxford University Press, 2003), pp. 216–222.
    45. Cohen, *Israel and the Bomb*, pp. 109–111; and Richelson, *Spying on the Bomb*, pp. 255–262.

Israel eventually allowed a few American inspections under tightly controlled conditions. The United States conducted inspections of Israel's nuclear facilities every year from 1961 through 1968, with the exception of 1963.[46] These inspections were too infrequent, however, to guarantee that nuclear fuel was not being diverted from the nuclear reactor toward other uses. Moreover, the visits were tightly controlled by an Israeli state bent on deceiving the United States about its technological progress. Israel permitted only rushed, one-day inspections of the site, barred U.S. officials from bringing their own scientific instruments, provided technical documentation of the facility's characteristics in Hebrew—a language that none of the inspectors could read—and prohibited the inspectors from bringing any of the documents out of Israel, citing security concerns. According to one source, rather than take American scientists to Dimona's actual control room, which would have given away the scale of Israel's plutonium production capability, Israeli scientists showed U.S. inspectors a mock-up control room provided by France that displayed data consistent with a small reactor. Further, to prevent the United States from discovering the underground plutonium reprocessing facility, the bank of elevators leading to the facility were walled off and plastered over during U.S. visits. Greenery was even planted outside the buildings in an attempt to camouflage any external signs of the reprocessing facility.

The subterfuge was successful. American inspectors consistently reported that they found no evidence of a plutonium reprocessing facility. In the words of author and intelligence expert Jeffrey Richelson, "Israel had developed a nuclear weapon under the noses and feet of U.S. inspectors and spies, both literally and figuratively."[47]

According to a common misperception, the United States was complicit in Israel's nuclear acquisition. This misperception may be fueled in part by unrealistic expectations about America's policy options, combined with confusion over the historical sequence of events.

It is imaginable that the United States could have taken more extreme measures, such as a preventive military strike against Israel's nuclear program, for example.[48] But this clearly would have been a disproportionate response. Rather, the United States attempted an entire range of more feasible nonproliferation policies to prevent nuclear proliferation in Israel.

---

46. On the U.S. inspections of Israel's nuclear facilities, see Cohen, *Israel and the Bomb*, pp. 175–194; Richelson, *Spying on the Bomb*, pp. 255–262; and Hersh, *Samson Option*, pp. 57, 111, 196.

47. Richelson, *Spying on the Bomb*, p. 262.

48. The United States could have also, for example, placed limits on private contributions from U.S. citizens to Israel.

As Israel pursued nuclear arms, consecutive U.S. administrations attempted to dissuade Israel of its nuclear ambitions. During the Kennedy and Johnson administrations, U.S. leaders placed diplomatic pressure on Israel's nuclear program; sought international cooperation from Egypt, the Soviet Union, and other states to develop multilateral measures to prevent nuclear proliferation in the Middle East; and forced a reluctant Israel to accept U.S. inspections of its nuclear facilities. Moreover, early U.S. decisions to provide conventional arms and security assurances to Israel were motivated in part by a desire to dissuade Israel from its nuclear course.[49] By providing Tel Aviv with the means to defend itself using conventional weapons, it was hoped that Israel would not need to resort to the nuclear option. The United States tied its transfers of military equipment to Israeli pledges to cooperate on the nuclear issue. Furthermore, as we saw above, U.S. leaders at the highest levels even threatened that the pursuit of nuclear weapons could jeopardize broader U.S. political support for Israel.

It is true that the intensity of U.S. pressure on Israel's nuclear program varied across presidential administrations.[50] Kennedy placed nuclear nonproliferation at the center of his administration's foreign policy agenda and consistently pushed within reasonable limits to stop Israel from acquiring nuclear weapons. In contrast, President Lyndon B. Johnson's focus on Great Society programs and his relative lack of interest in foreign policy made Israel's nuclear program a lower priority. Nevertheless, Johnson still preferred that Israel not acquire nuclear weapons and took active measures to dissuade Tel Aviv from its nuclear course.

In the end, of course, the United States was unable to sever France's nuclear assistance to Israel and to dissuade Israel from becoming a nuclear power. In 1962, the Dimona reactor went critical, and the plutonium reprocessing facility went online in 1965. It is widely believed that by 1967, Israel had constructed its first nuclear weapon, becoming a de facto member of the nuclear weapons club.[51]

Only after Israel acquired nuclear weapons did the American calculus change. When President Richard M. Nixon assumed office in 1969, he and National Security Advisor Henry Kissinger understood that Israel had already assembled a small nuclear arsenal and that persuading them to give it up would be highly unlikely.[52] They also foresaw the value of a U.S.-Israel alliance as a counterweight

---

49. See for example, Douglas Little, "The Making of a Special Relationship," *International Journal of Middle East Studies*, Vol. 25, No. 4 (November 1993), pp. 563–585; Cohen, *Israel and the Bomb*; and Hersh, *Samson Option*.

50. On policy differences between Kennedy and Johnson on nuclear proliferation in Israel, see for example Cohen, *Israel and the Bomb*; and Bass, *Support Any Friend*, pp. 186–238.

51. See, for example, Richelson, *Spying on the Bomb*, p. 242; and Cohen, *Israel and the Bomb*, p. 273.

52. On foreign policy in the Nixon administration, see, for example, Dallek, *Nixon and Kissinger*; and Fredrik Logevall and Andrew Preston, eds., *Nixon in the World* (New York: Oxford University Press, 2008).

to Soviet influence among the Arab states in the region. For these reasons, Nixon and Kissinger decided to engage a nuclear-armed Israel as a strategic partner and accepted Israel's policy of nuclear ambiguity.

By this point, however, the nuclear stage had already been set. Despite U.S. resistance, and because of large-scale French assistance, Israel had already become a nuclear power.

Explaining French-Israeli Nuclear Cooperation

In this section, I will use the strategic theory of nuclear proliferation to explain French-Israeli nuclear cooperation and U.S. opposition to nuclear proliferation in Israel. This section will demonstrate in detail how the observed pattern of sensitive nuclear assistance can be understood through a comparison of the strategic positions of the United States and France.

U.S. Strategic Position

From 1959 to 1965, the United States was one of the world's superpowers, and it enjoyed the ability to project conventional military power over Israel. As a power-projecting state, the United States feared that the spread of nuclear weapons to Israel would undermine its strategic position in the Middle East. Although there was no conceivable situation in which Washington would have wanted to attack Israel, U.S. officials feared that nuclear proliferation in Israel would impose a number of strategic costs on the United States. Furthermore, the United States did not share a rivalry with any of Israel's regional adversaries and did not have a strategic incentive to help Israel acquire nuclear weapons in order to constrain enemy states. Thus, the United States refused to provide Israel with sensitive nuclear assistance and instead tried to block French-Israeli nuclear cooperation.

U.S. POWER PROJECTION

The United States enjoyed the ability to project conventional military power over Israel during this time period. As the reader will recall from chapter 1, a state has a power-projection capability over another state, as I define it, if it has the ability to fight a full-scale conventional military war on that state's territory. Of course, it is unlikely that the United States would have wanted to invade Israel in this period. Still, as we saw in chapter 1, power-projecting states incur costs as nuclear weapons spread, whether the new nuclear weapon state is a friend or a foe. So while the United States might not have intended to invade Israel in this time period, an analysis of its ability to fight a full-scale conventional war on Israeli soil provides us with our best proxy measure of its ability to project military power over Israel.

Gauging the U.S. ability to invade Israel requires some speculation, since the United States did not actually attempt to invade Israel in this period. Nevertheless, an analysis of a number of factors strongly suggests that the United States could have put boots on the ground in a military contingency in Israel. These factors include vast U.S. military superiority over Israel using standard quantitative indicators of military power, U.S. force posture in the region, previous U.S. successes in launching amphibious invasions of hostile states in distant regions, and a proven ability to operate its military forces in the Middle East.

According to an accounting of standard quantitative indicators of military power, the United States was much more powerful than Israel in this time period. The United States greatly outspent Israel on defense matters. In the late 1950s and early 1960s, The United States spent about $50 billion annually on its military, dwarfing Israel's $130 million annual military budget.[53] Further, the United States maintained greater numbers of military personnel in uniform. The United States employed 2.5 million men in uniform ready for combat, compared to 65 thousand uniformed combat personnel in Israel. U.S. industrial production that could be mobilized for war was also vastly superior to that available in Israel. The United States produced 90 million tons of steel annually, while Israel produced only 80 thousand tons of steel each year. In sum, from 1959 to 1965, the years in which French-Israeli nuclear cooperation occurred, the United States possessed a massive 22 percent of global military power compared to the less than 1 percent of worldwide military power held by Israel.

U.S. military dominance over Israel in this time period is just as striking when U.S. power is discounted by the distance from Washington, D.C., to Tel Aviv. Standard military analysis adjusts a state's military power downward as the location of a military engagement extends away from a state's homeland. The vast power differential between these two states holds after discounting U.S. power by the 5,900 miles from Washington, D.C., to Tel Aviv.[54] In fact, the relative power differential between the United States and Israel, even after using a standard method for discounting by distance, places the dyad in the 90th percentile of the most one-sided of all the capable nuclear supplier–potential nuclear recipient dyads in the international system from 1951 to 2000.

---

53. All numbers on military power are taken from Correlates of War Project, *National Material Capabilities Data Documentation*, version 3.0, May 2005.
54. This method for discounting by distance was developed by Bueno de Mesquita, *War Trap*, and is further described in chapter 2.

Not only did U.S. military power vastly outstrip Israel's military might in terms of raw numbers, but the United States also possessed a force posture that would have allowed it to fight a full-scale conventional war on Israeli territory. The United States, with a few minor exceptions, did not maintain military outposts in the Middle East or North Africa during the Cold War.[55] The lack of military bases in the region would have prevented the United States from moving ground forces into Israel by land.

The United States did, however, maintain a strong naval presence in the Mediterranean. The U.S. Navy's Sixth Fleet contained a carrier battle group consisting of a minimum of one, and sometimes multiple, aircraft carriers supported by cruisers and destroyers.[56] The carriers hosted air wings, consisting of attack, fighter, anti-submarine, and patrol aircraft. The Sixth Fleet also maintained an amphibious assault force with assault ships and embarked landing craft, a Marine Expeditionary Unit, logistics units, and other forces. The Sixth Fleet was supported by American military bases throughout the Mediterranean including naval and air bases in France (until 1966), Spain, Italy, Greece, and Turkey. U.S. air bases in the region gave the United States an air presence across the entire Mediterranean.

With these naval and air forces, the United States almost certainly could have launched an amphibious invasion of Israeli territory in the face of a hostile defense.[57] The U.S. Marine Corps had perfected a method for conducting amphibious invasions. According to this method, air forces and naval aircraft establish air superiority and assert control over local sea areas. After establishing air and sea dominance and choosing a landing site, two-thirds of the designated amphibious forces storm ashore in landing craft, while the remainder is dropped by helicopter behind enemy lines. The helicopter-landed troops link up with the marines onshore. If the invasion plan does not allow for the capture of a port, an airfield is constructed on the beachhead and, in time, a temporary port. These facilities are used to bring ground troops ashore and to provide a re-supply route, positioning the United States to launch a full-scale military offensive.

The U.S. record for performing this type of operation in many different theaters in the face of hostile defenses strongly suggests that the United States could have recreated this feat in Israel during this time period. The Marine

---

55. The exceptions are two communications stations off the coast of Morocco, listening posts in Iran prior to the fall of the Shah in 1979, and some naval berthing privileges in Bahrain. National Defense University, "U.S. Military Bases in the Cold War."

56. On the capabilities of the U.S. Navy during the Cold War, see, for example, Robert W. Love Jr., *History of the U.S. Navy, 1942–1991* (Harrisburg, PA: Stackpole Books, 1992).

57. For a detailed discussion of amphibious warfare, see James F. Dunnigan, *How to Make War* (New York: Harper, 2003), 4th ed., pp. 284–298.

Corps mastered this method of amphibious invasion during World War II. In fact, every major offensive campaign staged by the United States in World War II began with an amphibious invasion.[58] These offensives included North Africa in November 1942, Sicily and Italy in July and September 1943, and Normandy and southern France in June and September 1944. The Marines also executed the U.S. strategy of "island-hopping" in the Pacific theater against the Japanese empire, fighting amphibious battles in Guadalcanal, Tarawa, Saipan, Iwo Jima, and Okinawa. The United States Marines also conducted successful amphibious invasions after World War II. Most notably, from a small toehold in Pusan, South Korea, the Marines captured Inchon, North Korea to turn the tide of the Korean War in September 1950.

The United States also demonstrated this type of force-projection capability in the Middle East in the period just prior to French-Israeli nuclear cooperation. The Eisenhower Doctrine, announced in 1957, was designed to limit Soviet influence in the Middle East. Eisenhower promised to defend the independence of the countries of the Middle East and offered to commit U.S. armed forces to protect any nation that was threatened by international Communism.[59] In April 1957, an aborted coup and the threat of further instability brought the United States military to the defense of Jordan's King Hussein. Vice Admiral Charles R. Brown ordered the carrier *Lake Champlain*, the battleship *Wisconsin*, two cruisers, and twenty-four destroyers to the eastern Mediterranean. Meanwhile, the carrier *Forrestal* and an amphibious task group carrying 1,800 Marines were stationed in Lebanese waters. President Eisenhower was prepared to send troops to Jordan if necessary, but Hussein managed to salvage his regime before the United States intervened.

In the following year, the United States put boots on the ground in order to influence the outcome of a conflict in Lebanon, a state on Israel's border. In 1958, a civil war broke out in Lebanon between the country's minority Christian and majority Muslim populations.[60] When Syria, which was backed by the Soviets, threatened to intervene in the conflict, the United States prepared for an invasion of Lebanon to protect U.S. interests. Under the leadership of Brown, the Sixth Fleet sent the heavy carrier *Saratoga*, the attack carriers *Essex* and *Wasp*, two cruisers, several divisions of destroyers, and amphibious assault teams to anchor off the coast of Cyprus. When on July 14, 1958, a coup d'état brought

---

58. For a discussion of the development of amphibious invasion tactics in the U.S. Navy, see Holland M. Smith, "The Development of Amphibious Tactics in the U.S. Navy," *Occasional Paper* (Washington, D.C.: History and Museums Division, U.S. Marine Corps, 1992), p. 2.

59. On U.S. foreign policy in the Middle East during the Eisenhower administration, see, for example, Roby C. Barrett, *The Greater Middle East and the Cold War* (New York: I.B. Tauris, 2004).

60. Ibid., pp. 420–428.

a pro-Soviet regime to power in Iraq, the United States and Lebanese President Camille Chamoun feared that Lebanon may be next. Eisenhower decided to send in the Marines. By the afternoon of the following day, July 15, 1958, four companies of U.S. Marines were on the ground in Lebanon. The United States retained its military presence in Lebanon until October 25, 1958.

In sum, the United States, through its superior military power and a force posture that included advanced amphibious invasion capabilities, had the ability to project military power over Israel and the wider Middle East region during this time period. The U.S. military even used an amphibious approach to occupy Lebanon, Israel's immediate neighbor to the north, in 1958. Had the United States needed to deploy troops to Israel, it almost certainly could have done so, either by bringing them overland from Lebanon, where the United States had proven that it could easily establish a foothold, or through a direct amphibious invasion of Israel itself.

Because the United States had the ability to project power over Israel, it had reason to worry that nuclear proliferation in Israel would undermine America's strategic position in the region. U.S. policymakers foresaw a number of negative strategic consequences of nuclear proliferation in Israel. U.S. officials feared that a nuclear armed Israel could constrain American military freedom of action in the region; force the United States to intervene in costly conventional conflicts between Israel and its neighbors; invite Soviet intervention in the region, potentially resulting in superpower war; render Israel less subject to U.S. influence; and set off further proliferation in the region, compounding the above-mentioned strategic costs.

The United States feared nuclear proliferation in Israel partly because it believed that nuclear weapons in the Middle East would constrain the U.S. military's freedom of action in the region. Since World War II, the United States had enjoyed the ability to influence regional politics with American military power, just as it had done in the 1958 Lebanese civil war. If states in the region acquired nuclear weapons, the ability of the United States to control the outcomes of conflict in the region would be impaired. This concern was reflected in a State Department working group report to Secretary of State Dean Rusk about the effects of nuclear proliferation in Israel and the United Arab Republic (a short-lived union of Syria and Egypt). In the State Department's estimation: "As programs developing sophisticated weapons come to fruition, the ability of the U.S. to control any hostilities which might occur between Israel and the United Arab Republic will decrease."[61] The report continued to advise a diplomatic initiative

---

61. Department of State, Memorandum, "Arms Limitations in the Middle East," May 14, 1963 in FRUS, 1961–1963, vol. 18, pp. 529–535, as cited in Cohen, *Israel and the Bomb*, p. 125.

to disarm both sides of their strategic forces because this would give the United States "more freedom of action to pursue unilateral means" to achieve its objectives in the region. In sum, the Gilpatric Committee report concluded that U.S. "military influence would wane" if Israel and other states acquired nuclear weapons.[62]

American officials further believed that nuclear proliferation in Israel could destabilize the region, potentially drawing the United States into a costly conflict. In a 1963 NIE, Sherman Kent wrote that with nuclear weapons, "Israel's policy toward its neighbors would become more rather than less tough...it would probably feel freer than it does now to take vigorous retaliatory action against border harassments when they did occur."[63] President Kennedy feared the potentially destabilizing effects of nuclear proliferation on the Middle East. In a letter to Ben-Gurion, Kennedy wrote that "development of such [nuclear] weapons would dangerously threaten the stability of the area."[64] The possibility that nuclear proliferation would destabilize the Middle East was a constant fear across administrations. In a memo to President Johnson, McGeorge Bundy warned, "President Nasser had indicated that the acquisition of a nuclear weapons capability by Israel would be cause for war no matter how suicidal for the Arabs."[65] In another memo to Johnson, NSC aide Robert Komer said that the acquisition of nuclear weapons by Israel "might spark Nasser into a foolish preemptive move." The Americans had good reason to fear conventional conflict arising out of Israel's nuclear weapons program. According to one report, Nasser himself, referring to Israeli's nuclear weapons program, threatened to "destroy the base of aggression, even at the price of four million casualties."[66]

Beyond the immediate effects of regional instability, the United States feared that nuclear proliferation in Israel would complicate the U.S.-Soviet relationship and could even set off a spiral toward superpower war. In the 1963 NIE, Kent predicted that the Arab reaction to Israel's nuclear weapons would be "profound dismay and frustration," and "among the principal targets of Arab resentment would be the U.S." The Arabs would turn to the Soviet Union, who would "win friends and influence in the Arab world."[67] Others worried that the Soviet Union, facing a mirror-image strategic situation, might also feel compelled to intervene

---

62. "A Report to the President by the Committee on Nuclear Proliferation," 21 January 1965, NSF, Box 5, LBJL, as cited in Cohen, *Israel and the Bomb*, p. 324.
63. Central Intelligence Agency, "Consequences of Israeli Acquisition of Nuclear Capability."
64. Kennedy, Letter to Ben Gurion.
65. Memorandum for McGeorge Bundy, "Subject: Need to Reassure President Nasser on the Peaceful Nature of the Dimona Reactor," n.d., Department of State, Center for Foreign Policy Files, RG 59, CFPF, 1964–1966, Box 3068, USNA, as cited in Cohen, *Israel and the Bomb*, p. 197.
66. Hersh, *Samson Option*, p. 108.
67. Central Intelligence Agency, "Consequences of Israeli Acquisition of Nuclear Capability."

militarily in the Middle East. This could lead to clashing superpower forces and, perhaps, superpower war. In a 1963 NIE, the consensus position of the United States intelligence community was that the impact of nuclear proliferation in the Middle East would be "the possibility that hostilities arising out of existing or future controversies could escalate into a confrontation involving the major powers."[68] According to one source, President Johnson also believed that a nuclear Israel meant increased Soviet involvement in the Middle East and perhaps superpower war.[69]

U.S. officials were also concerned that nuclear proliferation would reduce U.S. influence over Israel. Some U.S. officials speculated that if Israel acquired nuclear weapons, Israel would not only be less subject to American influence, it might also be in a position to force the United States to intervene on its behalf in a crisis.[70] On March 6, 1963, the head of the Office of National Estimates at the CIA, Sherman Kent, issued an eight-page memorandum entitled "Consequences of Israeli Acquisition of Nuclear Capability." According to Kent, a nuclear Israel "would use all its means at its command to persuade the U.S. to acquiescence in and even to support its possession of nuclear capability."[71] Some officials speculated that Israel could decide to aim its bomb "against the Americans, not to launch against America, but to say 'If you don't want to help us in a critical situation, we will require you to help us. Otherwise, we will use our nuclear bombs.'"[72]

U.S. policymakers further feared that the nuclearization of Israel would lead to more nuclear proliferation in the region, compounding the strategic costs detailed above. Specifically, U.S. officials worried that if Israel acquired nuclear weapons, the Soviet Union would station Soviet weapons on Arab soil or provide the Arab states with sensitive nuclear assistance. In a January 30, 1961, memorandum to President Kennedy, Secretary of State Dean Rusk warned, "Israel's acquisition of nuclear weapons would have grave repercussions in the Middle East, not the least of which might be the probable stationing of Soviet nuclear weapons on the soil of Israel's embittered Arab neighbors."[73] Kennedy apparently believed Rusk's forecast and decided to communicate these concerns directly to the Israeli leadership. In a letter to Ben-Gurion on May 17, 1961, President Kennedy

---

68. NIE, Number 4–63, June 28, 1963.
69. Hersh, *Samson Option*, p. 128.
70. For a theoretical discussion of entrapment see Glenn H. Snyder, "The Security Dilemma in Alliance Politics," *World Politics,* Vol. 36, No. 4 (July 1984), pp. 461–495.
71. Central Intelligence Agency, "Consequences of Israeli Acquisition of Nuclear Capability."
72. Hersh, *Samson Option*, p. 40.
73. Memorandum, Secretary of State Dean Rusk to President John F. Kennedy, "Subject: Israel's Atomic Nuclear Activities," January 30, 1961 CFSD (Central Files State Department), 884A.1901.1/1 Box 3061, USNA, as cited in Cohen, *Israel and the Bomb*, p. 102.

wrote, "It is difficult to imagine that the Arabs would refrain from turning to the Soviet Union for assistance if Israel were to develop nuclear weapons capability, what with all the consequences this would hold."[74]

Given the vast strategic costs that the United States would potentially suffer if Israel were to acquire nuclear weapons, it should come as no surprise that the United States vigorously opposed nuclear proliferation in Israel. According to Glen Seaborg, chairman of the AEC, nuclear proliferation in Israel was "Kennedy's private nightmare."[75] In communications with Israel, the United States was utterly unambiguous in its opposition to nuclear proliferation. Ogden Reid, U.S. Ambassador to Israel, explained to Israeli officials, "USG [U.S. government] policy is unequivocally opposed to proliferation of nuclear weapons capabilities."[76] Lewis G. Jones, U.S. assistant secretary of state for Near Eastern and South Asian affairs, told Avraham Harmon, Israeli Ambassador to the United States, that the "proliferation of nuclear weapons was absolutely anathema to the United States."[77] Secretary of State Dean Rusk wrote to Israeli Foreign Minister Abba Eban: "Israel should expect the U.S. to be extremely clear and utterly harsh on the matter of non-proliferation."[78] President Kennedy also told Israeli Prime Minister Levi Eshkol and Peres that "the United States was against proliferation of nuclear weapons in the world and she was certainly against their proliferation in the Middle East."[79] Washington's staunch nonproliferation stance was perhaps expressed most forcefully by President Johnson, who told Eshkol that the United States is "violently against nuclear proliferation."[80]

In sum, the United States enjoyed the ability to project military power over Israel, and this power-projection capability led U.S. officials to assess that nuclear proliferation in Israel would impose a number of strategic costs on the United States. For this reason, the United States opposed nuclear proliferation in Israel and refused to provide Israel with sensitive nuclear assistance.

---

74. Deptel 835 (Tel Aviv), Rusk to Barbour, 18 May 1963, POF, Israel, Box 119A, JFKL, as cited in Cohen, *Israel and the Bomb*, p. 128.

75. Glenn T. Seaborg with Benjamin S. Loeb, *Stemming the Tide* (Lexington, MA: Lexington Books, 1987), p. 249.

76. Telegram, Department of State to American embassy Tel Aviv, December 31, 1960 (Section 1), International File, Box 8: Israel, DDEL, as cited in Cohen, *Israel and the Bomb*, p. 93.

77. Department of State, *Foreign Relations of the United States, 1961–63: Near East*, Vol. 17 (Washington, D.C.: U.S. Government Printing Office, 1995), 9–10, as cited in Cohen, *Israel and the Bomb*, p. 103.

78. Deptel (652), Rusk to Barbor, February 10, 1966, RG 59, CFPF, 1964-66, Box 2356, USNA, as cited in Cohen, *Israel and the Bomb*, p. 211.

79. Peres, *David's Sling*, p. 103.

80. Memorandum of Conversation, Johnson, Feldman, Eshkol, and Harmon—First Meeting, June 1, 1964, ISA, FMRG, 3504/17, as cited in Cohen, *Israel and the Bomb*, p. 203.

U.S. COMMON ENEMIES

In the time period under study, the United States was not in a state of enmity with Egypt or any of Israel's other regional adversaries. Thus, the United States did not have a strategic incentive to provide Israel with sensitive nuclear assistance to constrain a common enemy.

From 1959 to 1965, the U.S.-Egyptian relationship was a delicate one, but according to the definition of a rivalry established in chapter 2, the United States was not a rival of Egypt, Israel's principal nemesis. Instead, prior to 1967, the United States was guided by a policy of neutrality in the Middle East, favoring neither Israel nor the Arab states.[81] As part of this policy of neutrality, the United States maintained a strict arms embargo on the Middle East. In 1950, the United States, along with France and Great Britain, signed a tripartite agreement in which the three parties pledged not to change the political-military status quo in the region. The agreement included a provision that prohibited the signatories from transferring conventional arms to the region.[82] France broke with the treaty in 1955 when it began to supply Israel with conventional military hardware in response to the Czech-Egyptian arms deal. In contrast, President Eisenhower maintained the embargo on Israel during the eight years of his presidency. As Eisenhower explained at the time, "We are not prepared to become the major supplier of arms to the Middle East."[83]

The lack of enmity between the United States and the Arab states in this period is further demonstrated by the U.S. response to the 1956 Suez Crisis. During the Suez Crisis, the United States did not side with Israel against Egypt. Instead, as part of Washington's policy of Middle Eastern neutrality, the United States backed Egypt against what President Eisenhower viewed to be unprovoked aggression on the part of France, Britain, and Israel. Even when the Soviet Union, the principal rival of the United States, intervened on Egypt's behalf and threatened Israel, the United States refused to come to Israel's defense. Instead, President Eisenhower joined Egypt and the Soviet Union in calling for an immediate ceasefire and demanding that Israel withdraw its troops from Egyptian soil.

In fact, Egypt and the United States even worked together on the common cause of preventing nuclear proliferation in Israel. Both states were concerned about the strategic implications of nuclear weapons in Tel Aviv and occasionally

---

81. On U.S. policy in the Middle East in this period, see, for example, Steven L. Spiegel, *The Other Arab-Israeli Conflict* (Chicago: University of Chicago Press, 1985); Jon B. Alterman, *Egypt and American Foreign Assistance, 1952–1956* (New York: Palgrave Macmillan, 2002); David Lesch, ed., *The Middle East and the United States* (New York: Westview Press, 2006); and Bass, *Support Any Friend*.

82. For more detail on Israel's search for arms in the West, see, for example, Peres, *David's Sling*, pp. 31–65.

83. Dwight D. Eisenhower, as cited in Peres, *David's Sling*, p. 37.

joined forces in an attempt to dissuade Israel from its nuclear course. The United States shared information with Nasser about Israel's nuclear program, met with Nasser to discuss ways to prevent nuclear proliferation in the Middle East, and attempted to negotiate a nuclear nonproliferation agreement between Egypt and Israel.[84]

Of course, this is not to say that the United States and Egypt were blessed with a harmonious relationship. They were not. There were many tensions between Cairo and Washington in this period. In 1956, for example, the United States reneged on a pledge to help President Nasser fund the Aswan Dam project, citing concerns about Nasser's ties with the Soviet Union. Neither would it be reasonable to deny that, over the course of this period, U.S.-Israeli relations began to warm. In the late Kennedy and early Johnson years, the United States lifted the arms embargo on Israel, and, following the 1967 Arab-Israeli War, the United States and Israel forged the close strategic relationship that exists to this day.[85]

Nevertheless, the United States and Egypt shared a functioning, if sometimes complicated, political relationship in this period and were not pitted against each other in a hostile military rivalry. Because the United States was not threatened by Nasser's Egypt or Israel's other regional adversaries, U.S. policymakers did not foresee any strategic benefit to constraining the military power of Arab states. According to the available evidence, U.S. officials never once considered that nuclear weapons in Israel could potentially benefit the United States by imposing strategic costs on Egypt or any of Israel's other regional rivals. Unlike France, the United States was not encouraged by an "enemy of my enemy is my customer" logic to supply sensitive nuclear materials and technology to Israel.

U.S. SUPERPOWER DEPENDENCE

During this time period, the United States was a global superpower. For this reason, the United States was threatened by nuclear proliferation in Israel and went on the offensive to prevent French-Israeli nuclear cooperation.

Since 1945, the United States has been a superpower with the ability to project power over the entire international system. As a reminder, Hypothesis 3 states that superpowers, because they have the ability to project power over the entire international system, are threatened by nuclear proliferation anywhere. For this reason, superpowers are not only unlikely to provide sensitive nuclear assistance, they are likely to take active measures designed to prevent the spread of nuclear weapons.

---

84. See for example Cohen, *Israel and the Bomb*; and Bass, *Support Any Friend*.
85. On the U.S.-Israeli strategic relationship, see Mearsheimer and Walt, *Israel Lobby and U.S. Foreign Policy*; and Herbert Drucks, *The Uncertain Alliance* (New York: Greenwood Press, 2001).

As we would expect from the strategic theory of nuclear proliferation, the United States, a superpower, took a number of steps aimed at halting French-Israeli nuclear cooperation and dissuading Israel from developing nuclear weapons.[86] The United States deployed intelligence assets to uncover the full extent of French-Israeli nuclear cooperation and Israeli nuclear development. Washington applied diplomatic pressure on Paris and Tel Aviv in an effort to suppress nuclear-related activity. The United States tied the provision of military aid to Israel's nuclear nonproliferation pledges. Further, the United States imposed a system of inspections on Israel to monitor and deter sensitive nuclear transactions and development. The possibility of nuclear proliferation in Israel led Washington (and Moscow) to consider more carefully an international treaty banning nuclear weapons possession. Finally, the United States even went so far as to threaten that nuclear proliferation in Israel could jeopardize U.S. support for Israel.

Perhaps one could argue that the United States was not truly willing to play hardball with Israel and could have done more to prevent sensitive nuclear cooperation between France and Israel. Still, there is no question that the United States was opposed to nuclear proliferation in Israel and, consistent with the expectations of Hypothesis 3, took a range of practicable policy steps in an attempt to stop it.

Shadow Case: The Soviet Union and Israel's Nuclear Program. The United States was not the only superpower opposed to nuclear proliferation in Israel. Though not the main focus of this chapter, a brief analysis of the Soviet Union's position on Israel's nuclear program provides further support for the strategic theory of nuclear proliferation. The Soviet Union was another superpower with global force-projection capabilities. Like the United States, Moscow was threatened by nuclear proliferation in Israel. Indeed, in a mirror image of Washington's fears, Moscow was concerned that a nuclear Israel would constrain Moscow's ability to project conventional military power in the Middle East, reduce the effectiveness of Moscow's coercive diplomacy on behalf of the Arab states and against Israel, cause instability in the region that could entangle the Soviet Union, distract an inordinate share of Moscow's strategic attention, and encourage further nuclear proliferation in the Middle East.[87] For example, the USSR Ministry of Foreign Affairs notified the Soviet embassies in Egypt and Israel, "The establishment of nuclear weapons production in Israel will make the situation...even

---

86. On U.S. efforts to prevent nuclear proliferation in Israel, see, for example, Cohen, *Israel and the Bomb;* and Bass, *Support Any Friend.*

87. See, for example, Ginor and Remez, *Foxbats over Dimona.*

more unstable, and is liable to trigger a serious conflict that can spill over the borders of the region."[88] Like the United States, the Soviet Union did not sit idly by while Israel developed nuclear weapons. Instead, the Soviet Union took actions designed to stop Israel's nuclear program. According to Ginor and Remez, the Soviet Union's concern about the strategic effects of nuclear proliferation in Israel led the Kremlin to develop military plans for a preventive strike on Israel's nuclear reactor at Dimona. Though their conclusions are controversial among some experts, the authors present evidence that suggests that Moscow's intense desire to destroy Israel's nuclear infrastructure and prevent nuclear proliferation in Israel may have been a contributing cause of the 1967 Arab-Israeli War.[89] Moscow may have even issued orders to Soviet military commanders to attack Israel's nuclear programs if certain contingencies were met in the 1967 war. Like the United States, the Soviet Union was a superpower intent on stopping the spread of nuclear weapons in the Middle East.

In sum, an analysis of the refusal of the United States to provide sensitive nuclear assistance to Israel strongly supports the strategic theory of nuclear proliferation. Supporting Hypothesis 1, United States had the ability to project military power over Israel, and U.S. officials feared that nuclear proliferation in Israel could constrain America's conventional military freedom of action. Consistent with Hypothesis 2, the United States and Israel did not share a common enemy, and U.S. officials did not foresee any strategic advantages that would come from nuclear proliferation in Israel potentially constraining Israel's enemies. And, the United States, a superpower, feared the strategic consequences of nuclear proliferation in Israel and took action to prevent nuclear cooperation between France and Israel, bolstering Hypothesis 3.

France's Strategic Position

France lacked the military capability to project conventional military force over Israel in this time period. For this reason, and unlike their American counterparts, French foreign policymakers did not foresee any grave strategic consequences that could result from nuclear proliferation to Israel. Furthermore, French officials actually foresaw strategic benefits to a nuclear-armed Israel. France was motivated to help Israel acquire nuclear weapons in order to constrain their shared rival: Egypt. France was able to continue supplying sensitive nuclear assistance, despite U.S. pressure, because France did not rely on the U.S. to provide for its security.

---

88. As quoted in ibid., p. 34.
89. Ibid.

FRANCE'S POWER PROJECTION

France lacked the ability to project power over Israel. While France clearly had the ability to bring air power to bear in the region, Paris lacked the ability to fight a full-scale conventional ground war in the Middle East. Therefore, French officials did not assess that nuclear proliferation in Israel would impose significant strategic costs on France. Thus, France could provide sensitive nuclear assistance to Israel without constraining its own military freedom of action.

As in the U.S. case, it is unlikely that France would have wanted to invade Israel in this period, but as we have seen, power-projecting states fear nuclear proliferation even to friendly states. Therefore, an analysis of France's ability to invade Israel provides us with our best measurement of the key independent variable, France's ability to project power over Israel. Of course, it is impossible to know for certain whether France could have invaded Israel and fought a full-scale conventional war in this period. Nevertheless, a number of factors demonstrate that France lacked an effective ability to project power over Israel. These factors include: rapidly declining French military power; a comparison of French and Israeli military power using standard quantitative indicators; France's force posture; and France's inability to bring military force to bear singlehandedly in other Middle Eastern conflicts, including the Suez Crisis, in this period.

France's global military presence drastically receded during World War II and in the years that followed. In the early twentieth century, France possessed a vast empire that was supported by a global military presence. During World War II, however, major military battles were fought on French territory, devastating France's economy, society, and military.[90] As France's military power waned, so too did its ability to maintain a global empire. By the time the sensitive nuclear cooperation between France and Israel began in 1959, France had relinquished control over much of its former empire.[91] In the Middle East, France granted Lebanon and Syria independence in the mid-1940s. In North Africa, Morocco and Tunisia achieved independence in 1956. Algeria launched a long and bloody war of independence beginning in 1954. In Southeast Asia, the French suffered a devastating military defeat at the battle of Dien Bien Phu, leading to a French withdrawal from Indochina in 1954. In Africa, France would lose Guinea in 1958, and Benin, Burkina Faso, Cameroon, Central African Republic, Chad, Congo-Brazzaville, Cote D'Ivoire, Gabon, Madagascar, Mali, Mauritania, Niger, Senegal,

---

90. On World War II's effects on France, see, for example, Thomas R. Christofferson with Michael S. Christofferson, *France during World War II* (New York: Fordham University Press, 2006); and Peter Davies, *France and the Second World War* (New York: Taylor and Francis, 2007).

91. On France's empire, see, for example, Frederick Quinn, *The French Overseas Empire* (New York: Praeger, 2001); and David B. Abernathy, *The Dynamics of Global Dominance* (New Haven: Yale University Press, 2002).

and Togo in 1960. In sum, France's global empire had virtually vanished in the span of just two decades. By the time French-Israeli nuclear cooperation was proceeding apace in the early 1960s, France's major colonial possessions had been reduced to a war-torn Algeria, Djibouti, French Equatorial Guinea, and scattered islands in the Indian Ocean and Caribbean. The collapse of the empire was not only a consequence, but also a further cause, of French military decline. Deprived of foreign bases of operation, France's ability to project power in distant regions was reduced.

In terms of raw numbers, France's military was vastly superior to that of Israel.[92] France spent about five billion dollars per year on its military, maintained 585 thousand troops in uniform ready for combat, possessed a total population of 46 million people, and produced 17 million tons of steel per year. The comparable numbers for Israel were: $130 million per year in military spending; sixty thousand uniformed combat troops; a population of two million people; and eighty thousand tons of steel per year.

Taking into account the geographical distance between the two states, however, a very different picture emerges. After using a standard formula for discounting French military power by the 2,073 miles from Paris to Tel Aviv, France becomes weaker than Israel in relative terms.[93] That is, after adjusting for distance, France's aggregate military power is less than the aggregate military power of Israel. In a hypothetical war between France and Israel fought on Israeli soil, Israel would have enjoyed a slight numerical advantage.

A detailed military analysis of French defense posture adds to the evidence suggesting that France lacked the ability to project power over Israel in this period. French basing posture would have rendered a ground invasion of Israel impossible. France's military bases nearest to Israel were located in Algeria, Djibouti, and Senegal. A ground invasion staged from these bases would have required French forces to march great distances over rough terrain, through hostile states. It is implausible that France could have conducted such an operation.

Unable to approach Israel by land, France's only option would have been an amphibious invasion of Israel either from the Mediterranean Sea or the Red Sea. But, unlike the United States, France lacked the capabilities required for an amphibious invasion of this sort. Eighty-five percent of the French naval fleet was destroyed during World War II.[94] The French were slow to rebuild their navy after

---

92. All numbers on military power are taken from Correlates of War Project, *National Material Capabilities Data Documentation.*

93. This method for discounting by distance was developed by Bueno de Mesquita, *War Trap*, and is described in chapter 2.

94. Christofferson, *France during World War II*; and Davies, *France, and the Second World War.*

the war. French military planners did not foresee a contingency in which France would need to launch an amphibious invasion of a distant country and, accordingly, did not invest in these capabilities.

Instead, in the post–World War II era, the French military was designed to perform two primary functions: counterinsurgency and policing in the colonies and the defense of France's Eastern border against a potential Soviet invasion.[95] From 1945 to 1960, French military planners focused on shoring up the remnants of the French empire with capabilities designed to conduct policing and counterinsurgency operations in French colonial possessions and overseas departments. Thus, military investments were made primarily in manpower. Focusing on traditional counterinsurgency missions, the French military missed out on the military modernization that was occurring in the United States and other industrialized nations. According to Michel Martin, until 1960, the French military was "dominated by a huge land force ... [and a] technologically deprived air force and navy."[96]

In 1960, as the French Empire dissipated, French strategic planners explicitly rejected a global role for the French military. Instead, they turned their attention to defending France from a potential Soviet invasion.[97] For the remainder of the Cold War, French grand strategy rested on deterring a Soviet invasion with the threat of nuclear retaliation. Thus, military investments were disproportionately devoted to strategic nuclear forces. France's nuclear-oriented strategy led it to reduce investments in conventional capabilities and drastically decrease military spending across the board. The French navy fought for, and eventually received, its share of strategic nuclear forces. The navy operated nuclear submarines equipped with submarine-launched ballistic missiles and, in 1963, commissioned two new aircraft carriers, *Foch* and *Clemenceau*, with air wings comprised of nuclear-armed bombers. The new carriers expanded the range of France's strategic bombing capability, but did not provide France with an amphibious invasion capability. Unlike their American counterparts, the French aircraft carriers were part of fleets that did not include significant amphibious assault units. Instead, the French possessed only a few landing craft sold to them by the British at the end of World War II.[98]

In fact, France never developed a meaningful amphibious invasion capability. On a quantitative index scoring the amphibious invasion capabilities of the

---

95. On French defense policy during the Cold War, see, for example, Michel L. Martin, *Warriors to Managers* (Chapel Hill: University of North Carolina Press, 1981); and Sten Rynning, *Changing Military Doctrine* (London: Praeger, 2002).
96. Martin, *Warriors to Managers*, p. 38.
97. Martin, *Warriors to Managers*; and Rynning, *Changing Military Doctrine*.
98. Barker, *Suez*, p. 26.

countries of the world on a scale from zero to one thousand, a leading military analyst scored the United States a perfect one thousand.[99] The same analyst scored France a mere seventeen. In fact, France's meager amphibious invasion capabilities placed it behind such military lightweights as Thailand (32/1,000) and the Philippines (24/1,000).

The performance of the French military in the eastern Mediterranean in this period further demonstrates France's inability to project military power over Israel. The best evidence that France lacked a force-projection capability in the Middle East is the simple absence of a French military presence in the region after World War II. The French military did not intervene in many Middle Eastern crises in which it had a clear stake. For example, in the case of Lebanon, a former French colony, it was the United States, not France, that intervened in the 1958 civil war. Indeed, to the chagrin of many French officials, the United States did not even feel obliged to consult with France or to consider the incorporation of French military forces in the operation.[100] This lack of consideration was largely due to France's dearth of capabilities that could have been brought to bear in the crisis.

Some may wonder whether France's participation in the 1956 Suez Crisis proves that France possessed the ability to project military power in the Middle East in this time period. In fact, on the contrary, a detailed examination of the Suez Crisis demonstrates that France lacked the independent ability to project power in the Middle East. At Suez, France was completely dependent on British and Israeli basing and military forces. When President Nasser seized the Suez Canal, French leaders were eager to intervene to protect French economic and strategic interests, but France, acting alone, did not possess a viable military option. According to A. J. Baker, France did not have "the trained men available, aircraft, or amphibious means with which to transport them to Egypt in order to enforce their policies."[101]

The lack of a French force-projection capability in the region compelled the French to seek allies for a joint military operation. Recognizing France's international political position as well as the limits of French military power, French Prime Minister Guy Mollet understood that any Suez operation could only "take the form of joint action by the Western Allies."[102] The French quickly identified Israel and Britain as the proper partners for military action in Egypt.

---

99. Dunnigan, *How to Make War*, p. 293.
100. See, for example, Philip G. Cerny, *The Politics of Grandeur* (Cambridge: Cambridge University Press, 1980), p. 167.
101. Barker, *Suez*, p. 19.
102. Guy Mollet, as cited in Jean-Paul Cointet, "Guy Mollet, the French Government and the SFIO," in Troen and Shemesh, eds., *Suez-Sinai Crisis 1956*, p. 133.

Israel was well positioned to launch a cross-border ground invasion into the heart of Egypt. Reflecting France's interest in, and need for, Israeli ground forces in an operation against Egypt, Bourgès-Maunoury summoned Shimon Peres to his office the day after Nasser's seizure of the Suez Canal to ask "how long it would take the Israeli army to cross the Sinai Peninsula." Peres responded that "our army people estimated it would probably take from five to seven days." Realizing the valuable role that Israel could play in the operation, Bourgès-Maunoury immediately followed up by asking if Israel would "be prepared to take part in a tripartite military operation, in which Israel's specific role would be to cross the Sinai."[103]

France desired British participation in the Suez operation because Britain, unlike France, possessed military bases in the region and had the capability to lead an amphibious invasion of Egypt. The French dependence on British military capability during the Suez Crisis is illustrated by André Martin, a French military officer involved in the planning for Suez who explained that "for geographical and political reasons, France required a partner, and the only possible choice was Britain.... It was understood that without [the British bases in] Malta and Cyprus, we could do nothing, and we really wanted this war!"[104] Israeli participants also recall British basing as the primary reason that the French were eager to secure British participation. According to Peres, France sought British cooperation because "this gave France a European partner with a base in the Middle East (Cyprus)."[105] Britain and Israel could put boots on the ground in Egypt, but French foreign policymakers understood that France, without assistance from its allies, could not.

The conduct of the Suez Crisis further demonstrated France's inability to project power unilaterally in the region.[106] The combined Anglo-French operation to capture the Port of Said, code-named "Musketeer," had three major phases. An analysis of each stage of Musketeer reveals the inability of the French to project power unilaterally in the region. In the first phase, British and French aircraft conducted bombing raids to weaken Egyptian air forces and military targets.[107] The initial bombing campaign relied heavily on British basing and forces. The vast majority of the sorties were flown from British air bases in Cyprus. The remainder of the sorties were flown from British and French aircraft carriers in the Mediterranean Sea. The use of the carriers themselves, however, depended on

---

103. Peres, "Road to Sèvres," pp. 141–142, 122.
104. André Martin, "The Military and Political Contradictions of the Suez Affair," in Troen and Shemesh, eds., *Suez Crisis 1956*, p. 54.
105. Peres, *David's Sling*, p. 185.
106. For a detailed discussion of military operations during the Suez Crisis, see Barker, *Suez*.
107. On phase one of Operation Musketeer, see ibid., pp. 97–108.

British bases; the carriers steamed to Egyptian waters from the British naval base in Cyprus. For the carrier-based sorties, the British contributed five modern aircraft carriers, operating air wings consisting of modern jet bombers. The French contributed their entire carrier fleet: two light carriers left over from World War II, equipped with air wings of outdated, piston-engine planes.

Musketeer's second phase consisted of an aerial assault in which British and French paratroopers dropped into Egypt to capture key strategic points near the Port of Said. Again, this phase of the operation depended on British basing and forces.[108] The British 16th Independent Parachute Brigade and the French 10th Airborne Division provided the airborne spearhead. The paratroopers were flown into Egyptian airspace from British air bases in Cyprus. Six hundred British and five hundred French paratroopers participated in the assault. The British led the operation, taking the airfield at Gamil at the Port of Said. French paratroopers followed the British, landing near the twin bridges at Raswa that linked the Port of Said to the Suez Canal.

Musketeer's third and final phase was an amphibious invasion to capture the Port of Said. This phase was conducted almost entirely by British forces.[109] The assault began with the bombardment of the Port of Said from artillery and guns mounted on British and French naval forces in the Mediterranean. The bombardment, again, depended heavily on British basing and forces. The naval fleets steamed to Egypt's coast from British naval bases in Malta. One hundred British and thirty French warships participated in the assault. The bombardment prepared the way for the amphibious invasion. The British 3rd Commando Brigade formed the backbone of the seaborne assault. Two British commando units stormed ashore in assault landing craft. A third British commando unit followed in helicopters. The French, due to a lack of capability, did not participate directly in the amphibious assault.

The conclusion of the Suez crisis further demonstrated France's military impotence in the region. In the face of Soviet threats, France and Great Britain were forced to beat an embarrassing retreat. When Peres asked what France's response would be to a Soviet military intervention against Israel, French officials were willing but unable to offer military protection.

In short, without British and Israeli assistance, France would not have been able to participate fully in the Suez Crisis. At every stage of the operation, France depended on superior British basing, forces, and capabilities. Thinking counterfactually, it is possible to imagine that France may have been able to conduct a much more limited military campaign against Egypt without British assistance.

---

108. On phase two of Operation Musketeer, see ibid., pp. 109–129.
109. On phase three of Operation Musketeer, see ibid., pp. 143–156.

Without British help, French military forces may have been able to conduct a limited bombing campaign and may have been able to land some paratroopers on the ground in Egypt. Lacking bases in the region and amphibious capabilities, however, it is entirely implausible to imagine that France could have waged a full-scale, conventional, ground war in the region.

Since France lacked the ability to project conventional military power against Israel, France would have forfeited very little, if any, of its own military freedom of action if Israel were to acquire nuclear weapons. In other words, the nuclearization of Israel did not directly undermine France's strategic position.

For this reason, unlike in the American case, French officials did not assess that nuclear proliferation in Israel would constrain France's conventional military freedom of action. Indeed, in their analyses of the likely effects of nuclear proliferation in Israel, French officials did not identify any substantial strategic costs. As we saw above, French officials did worry about the possible international reaction to French-Israeli nuclear cooperation, but, according to the available evidence, French officials did not believe that nuclear proliferation in Israel would in any way impose strategic costs on France. Of course, the absence of evidence is not necessarily evidence of absence. Nevertheless, given the wealth of materials that are available, including personal statements by officials directly involved in the nuclear transfers at the highest levels of the French government, the lack of any indication that French officials were worried about the strategic consequences of nuclear proliferation in Israel is quite striking.

In sum, France, unlike the United States, lacked the ability to project power over Israel. For this reason, French officials did not assess that nuclear proliferation in Israel would constrain its own military freedom of action. Therefore, France could provide sensitive nuclear assistance to Israel without self-inflicting grave strategic wounds. Not only did nuclear proliferation in Israel not threaten France directly, French policymakers also identified a number of strategic benefits from nuclear proliferation in Israel.

FRANCE'S COMMON ENEMIES

In direct contrast to American calculations, French officials concluded that there were a number of direct strategic benefits that France could gain by helping Israel acquire nuclear weapons. France and Israel shared a common enemy: Egypt. By providing sensitive nuclear assistance to Israel, France hoped to impose strategic costs on this shared rival.

In the late 1950s, France's foremost strategic objective was constraining Nasser's military power in the Middle East. French officials believed that severing Egypt's support to the rebels in Algeria was a necessary precondition for defeating the Algerian insurgency. According to Peres, "Some [French] leaders, notably those responsible for defence matters, held that clipping Nasser's wings would limit his ambitions and impact on the Algerian front."[110] Jean Chauvel, France's Ambassador to Great Britain, stated that if France failed in its efforts to counter Nasser's growing power, "it would be impossible for France to pursue the struggle in Algeria."[111] The importance that the French government placed on unseating Nasser is reflected in the words of French Prime Minister Guy Mollet, who said of his leadership in the Suez Crisis, "I have only one regret: not to have been able to go all the way (and overthrow Nasser)."[112]

To constrain Egyptian power, France pursued a strategy of guaranteeing the security of Israel, Egypt's main regional rival, through transfers of military technology. The timing of French-Israeli military and nuclear agreements, always in direct response to an increase in the power or potential influence of Egypt, demonstrates the strategic nature of French-Israeli cooperation. The initial conventional arms agreements between the two sides followed immediately in the wake of the Czech-Egyptian arms deal in 1955. The first nuclear agreement between the two countries was a response to Nasser's decision to nationalize the Suez Canal in 1956. And, finally, the agreement for the provision of sensitive nuclear assistance occurred at the height of the Suez Crisis.

French officials knew that the nuclear transfers would be used for weaponization and had no illusions that these would be civilian transfers. Bertrand Goldschmidt later explained, referring to Israel's initial requests for nuclear assistance, "Although it was not mentioned, [the Israelis] implied that their goal was weapons production."[113] And, when asked by a French diplomat in 1963 if France was helping Israel to construct a nuclear bomb, Goldschmidt is reported to have replied with a laugh, "Not only did we take the girl when she was a virgin, but we made her pregnant."[114]

Finally, the recorded statements of French officials confirm that the sensitive nuclear assistance was designed to ensure Israel's security against threats from Egypt and its Soviet patron. At the height of the Suez Crisis, Pineau and General Paul Eli, chief of staff of the French army, said that they were willing to provide Israel with their most advanced nuclear technology to help guarantee Israeli security. According to Peres, at the fateful Paris meeting in November 1956, Pineau promised Peres that "they were ready to share with us everything that they had."[115] General Eli agreed. Referring to Israel's

---

110. Peres, *David's Sling*, p. 46.
111. Jean Chauvel, as cited in Louis and Owen, *Suez 1956*, p. 137.
112. Guy Mollet, as cited in Cointet, "Guy Mollet," p. 138.
113. Pinkus, "Atomic Power to Israel's Rescue," p. 119.
114. Hersh, *Samson Option*, p. 119.
115. Peres, "Road to Sèvres," p. 131.

request for sensitive nuclear assistance, Eli turned to his French colleagues and said, "We must give them this to guarantee their security, it is vital."[116] In later years, key French officials would recall the provision of sensitive nuclear assistance to Israel in similar terms. Describing his motivations in later years, Mollet said, "When my government came to power, Israel asked for French assistance; I did my duty as a democrat and a Frenchman by supplying this endangered country with the arms it needed to survive."[117] French Defense Minister Bourgès-Maunoury later explained, "I gave [the Israelis] the atom...so that Israel could face its enemies in the Middle East."[118] As Bourgès-Maunoury reportedly confided to Peres at the time, "France and Israel now faced similar challenges and similar foes, and we should co-operate openly—and quickly.... We should work together and we can."[119]

While we only have a few statements from French officials responsible for authorizing the transfers, their testaments, combined with other available evidence, strongly suggests that French nuclear assistance to Israel was part of a strategy to help Israel acquire nuclear weapons in order to constrain Egypt. The statements from French officials make it clear that the sensitive nuclear transfers were motivated by security concerns and not economic or other matters. This security motivation is further supported by the timing of the initial decision, at the height of the Suez Crisis. Finally, it is well documented in the historical record that the French-Israeli strategic partnership was the result of a shared desire to counter a shared enemy in the region. France and Israel both wanted to constrain Nasser's Egypt.

In hindsight, one may question the logic under which the French officials were operating. Proliferation optimists argue that the spread of nuclear weapons leads to international stability.[120] If French officials had conceived of the effects of nuclear proliferation in similar terms, they could have decided that nuclear proliferation in Israel would have been counterproductive to French interests. Nuclear weapons in Israel could have had the unwanted effect of inducing stability in the Egyptian-Israeli relationship, freeing up Nasser to pour all of his resources into the Algerian conflict, thereby only further complicating France's position. While this argument is logical, it is at odds with the empirical evidence. We have seen that, unlike many academics, policymakers are not proliferation optimists. French, Israeli, and Egyptian officials were united in

---

116. Pinkus, "Atomic Power to Israel's Rescue," p. 123.
117. Guy Mollet, *France Observateur*, August 10, 1956, as cited in Troen and Shemesh, eds., *Suez-Sinai Crisis 1956*, p. 131.
118. Maurice Bourgès-Maunoury, as cited in Pinkus, "Atomic Power to Israel's Rescue," p. 118.
119. Peres, *David's Sling*, p. 57.
120. See for example Waltz, "More May Be Better."

believing that nuclear proliferation in Israel would impose significant strategic costs on Egypt.

Indeed, the preponderance of evidence strongly suggests that France's sensitive nuclear assistance to Israel was part of a French strategy to ensure Israel's security in order to constrain Egypt, an enemy that both France and Israel shared.

FRANCE'S SUPERPOWER DEPENDENCE

France was able to provide sensitive nuclear assistance in spite of U.S. opposition because France was not dependent on superpower pressure. While France was exporting sensitive nuclear materials and technology to Israel, France possessed its own nuclear arsenal and did not depend on the superpowers to provide for its security. For this reason, France's security needs would not be seriously threatened if its relationship with the superpowers soured, so it could afford to provide sensitive nuclear assistance to Israel despite superpower pressure. France seriously considered U.S. objections to nuclear cooperation with Israel but continued with the sensitive nuclear transactions anyway.

In 1949, France became a founding member of NATO and found itself under the nuclear umbrella of the United States. But French officials soon decided that they could not afford to outsource their core security needs. De Gaulle's line about America's willingness to trade New York for Paris comes to mind. In the 1950s, France began to take the steps necessary to weaponize its civilian nuclear capability. Under the French Fourth Republic (1946–1958), French officials launched a number of research programs to expand and develop France's nuclear capabilities.[121] Officials in the Fourth Republic did not make an explicit decision to acquire nuclear weapons, but they gave France the latent nuclear capability that would allow future French leaders to go nuclear. When de Gaulle assumed power at the beginning of the Fifth Republic, he was eager to exercise the option bequeathed to him. In October 1958, de Gaulle declared, "Everybody knows that we now have the means of providing ourselves with nuclear weapons and the day is approaching when we, in our turn, will carry out tests."[122] De Gaulle kept his promise. In February 1960, France's security independence was assured when France tested its first nuclear weapon.

The security guarantee afforded by a national nuclear weapons arsenal allowed French foreign policymakers to pursue a more independent foreign policy. According to de Gaulle, independence would be the "essential goal" of French foreign policy. Rather than form a firm commitment with the Western alliance,

---

121. For the definitive source on nuclear policy in the French Fourth Republic see, Scheinman, *Atomic Energy Policy*.

122. Charles de Gaulle, as cited in Scheinman, *Atomic Energy Policy*, p. xvi.

France would, according to de Gaulle, "collaborate with the West and the East, constructing alliances with one side or the other as necessary, without ever accepting a position of dependence."[123]

De Gaulle's goal of independence manifested itself in an increasingly autonomous French security policy. In March 1959, France withdrew its Mediterranean fleet from NATO. In 1959, de Gaulle banned the stationing of foreign nuclear weapons on French soil, forcing the United States to remove over two hundred nuclear-armed aircraft from France. France removed its French Atlantic and Channel fleets from NATO command in 1962. The policy of military independence reached its apogee in 1966 when France completely withdrew from NATO's unified military command structure. NATO's headquarters were transferred from Paris to Belgium, and the United States was forced to abandon ten military bases on French soil. In short, from 1959 to 1965, the period in which France provided Israel with sensitive nuclear assistance, France enjoyed functional independence from superpower pressure.

Some may wonder whether France's independent foreign policy stance in this period was less the result of the structural conditions of superpower dependence identified here and more the result of an idiosyncratic leader. De Gaulle may have been a unique historical figure, but France's foreign policy of independence cannot be seen as the result of any single individual, not even de Gaulle. De Gaulle's preference for an independent French foreign policy cannot be divorced from the structural and material conditions that facilitated it. French concerns about overreliance on American security guarantees were rooted in the post–World War II settlement.[124] These concerns again flared up in the aftermath of the Suez Crisis, leading French officials to begin building the material basis for an independent French security policy. As Lawrence Scheinman has argued, the Gaullist foreign policy of independence would not have been possible if the groundwork had not been laid during the French Fourth Republic. Scheinman writes, "The decisions which led to the detonation of France's first atomic device at Reggane on February 13, 1960 were taken while General de Gaulle was in [a mid-career] retirement at Colombey-les-deux-Eglises."[125] In other words, it was not a Gaullist foreign policy that brought France to a position of independence. It was the structural and material basis for independence that permitted a Gaullist foreign policy.

That France did not rely on a superpower to provide for its own security allowed France to provide sensitive nuclear assistance to Israel despite U.S. objections. French officials seriously considered possible American and international

---

123. Kohl, *French Nuclear Diplomacy,* pp. 128, 127.
124. Cerny, *Politics of Grandeur,* p. 14.
125. Scheinman, *Atomic Energy Policy,* p. xii.

reactions to its sensitive nuclear assistance to Israel. As we saw above, de Gaulle was concerned about a possible international reaction to French nuclear assistance to Israel. But he was not overly concerned. Instead of canceling the sensitive nuclear transactions, de Gaulle handed the task of nuclear assistance to French firms, providing the French government with a degree of plausible deniability. As French nuclear scientist Jean-Francis Perrin later claimed, "We wanted to help Israel. The secret was kept because of the Americans."[126]

In the end, France decided to proceed with the sensitive nuclear transfers in the face of U.S. resistance. Though France carefully considered the international diplomatic costs of aiding the nuclear ambitions of another state, these were outweighed by the strategic benefits. According to one analyst of French-Israeli nuclear cooperation, "France was ready to risk harming its relations with the Arab world and the superpowers by extending assistance to Israel for establishing a nuclear infrastructure with an indisputable military capability."[127]

Shadow Cases: Argentina and Taiwan. We have seen that superpower independence was a contributing factor in France's decision to provide sensitive nuclear assistance to Israel. This claim rests on the counterfactual argument that had France been dependent on a superpower security guarantee, it would have paid more attention to the superpower's demands and would have been much less likely to participate in the sensitive nuclear transaction. In this section, I will briefly examine additional cases to demonstrate that states that are vulnerable to superpower pressure will tend to find that the costs of superpower displeasure outweigh the potential benefits to be gained from providing sensitive nuclear assistance.[128] For this reason, these states refrain from supplying sensitive nuclear materials and technologies. When superpower-dependent states attempt to provide or receive sensitive nuclear assistance, a superpower intervenes, and the dependent states cancel the transaction in the face of superpower pressure.

The effect of superpower dependence on sensitive nuclear assistance is generally unobservable. States that are vulnerable to superpower pressure look to the future and realize that attempts to provide sensitive nuclear assistance would likely result in an intervention from the superpower patron. Faced with the possibility of having their security guarantees revoked, the would-be nuclear providers understand that they would only relent in the face of superpower pressure and cancel the transaction. Given the futility of such an exercise and the potential costs of crossing a superpower patron, superpower dependent states generally

---

126. Pinkus, "Atomic Power to Israel's Rescue," p. 128.
127. Ibid., p. 131.
128. For a discussion of adding additional cases to one's study as a method for improving causal inference, see King, Keohane, and Verba, *Designing Social Inquiry,* p. 219.

refrain from attempting to provide sensitive nuclear assistance altogether. The logic of this argument is supported by the strong negative relationship between superpower dependence and sensitive nuclear assistance demonstrated in chapter 2. The effect of superpower dependence on sensitive nuclear assistance can also be seen in the fact that many of the capable nuclear suppliers that rely on a superpower security guarantee, or are themselves superpowers, have never provided sensitive nuclear assistance. The list of such states include Argentina, Belgium, Brazil, Japan, the Netherlands, Norway, and the United States.

The mechanisms inducing restraint in superpower-dependent states are observable in the occasional cases in which a superpower-dependent state has attempted to provide or receive sensitive nuclear assistance. In the case of proposed Argentinean-Libyan nuclear cooperation in 1985, a superpower-dependent state agreed to participate in a sensitive nuclear transaction but canceled the transfer in the face of superpower pressure. In another case, Taiwan halted a nuclear cooperation agreement with France in 1975, due to intense U.S. opposition. Although Taiwan was the recipient and not the supplier in this proposed deal, the details of the case help to illustrate the mechanisms by which superpower pressure influences sensitive nuclear transfers involving superpower-dependent states.

Argentina is superpower-dependent. It lacks a nuclear weapons arsenal and is in a defense pact with the United States. Argentina is a signatory of the 1948 Inter-American Treaty of Reciprocal Assistance, commonly known as the Rio Pact. The treaty was established by the United States at the beginning of the Cold War to bring the countries of the Western Hemisphere under the American security umbrella. According to Article III of the treaty, the member states "agree that an armed attack by any State against an American State shall be considered as an attack against all the American States and, consequently, each one of the said Contracting Parties undertakes to assist in meeting the attack."[129]

Since entering into the defense pact with the United States, Argentina's dependence on the United States has been reflected in its foreign policy doctrine. Martin Mullins describes Argentina's foreign policy as "an explicit commitment to accepting U.S. leadership."[130] Mullins argues that Argentinean officials concluded that Argentina could not "derive any benefit from conflicts with the United States. Thus, it was recommended that Argentina should...accept its dependent position."[131]

---

129. See the Inter-American Treaty of Reciprocal Assistance.
130. On Argentina's foreign policy, see, for example, Martin Mullins, *In the Shadow of Generals* (Burlington, VT: Ashgate, 2006), p. 62.
131. Martin Mullins, *In the Shadow of Generals* (Burlington, VT: Ashgate, 2006), pp. 62, 64.

Argentina's dependence on the United States influenced its nuclear export policy. In 1969 Argentina developed a laboratory-scale plutonium reprocessing capability, also giving it the ability to provide sensitive nuclear assistance. For over forty years, Argentina had the option of exporting sensitive nuclear materials and technology, but never exercised this option. This restraint is partly the result of latent U.S. pressure and the understanding that Washington would vigorously oppose any attempted nuclear exports from Buenos Aires. For this reason, Argentina did not even attempt to provide sensitive nuclear assistance. In 1985, however, Argentina tested the U.S. position on sensitive nuclear exports. In that year, Argentina entered into an agreement to send plutonium reprocessing facilities to Libya.[132] When the United States discovered the existence of the planned nuclear transfer it intervened and forced Argentina to cancel the sensitive nuclear export. Argentina caved under U.S. pressure and canceled the proposed transaction.

Taiwan is another state that was dependent on a U.S. security guarantee and canceled planned participation in a sensitive nuclear transaction under American pressure. The United States first extended a security guarantee to Taiwan (formally known as the Republic of China) on June 25, 1950, at the beginning of the Korean War. President Harry S. Truman publicly announced, "I have ordered the Seventh Fleet to prevent any attack on Formosa."[133] The security relationship between the two sides was formalized in 1954, when the United States and Taiwan signed a mutual defense treaty. The security guarantee lasted until 1979 when the U.S. Congress passed the Taiwan Relations Act (TRA). TRA pledges the United States to maintain military capabilities to ensure peace in the Western Pacific, but it does not require the United States to intervene on Taiwan's behalf in the event of an attack from mainland China.

In 1969, Taiwan requested plutonium reprocessing facilities from the United States, but the Nixon administration denied the request and barred U.S. firms from selling the key component parts necessary for plutonium reprocessing to Taiwan.[134] Like Israel, Taiwan, spurned by the United States, turned to France as a potential nuclear supplier. In 1973, France and Taiwan entered into a nuclear cooperation agreement. The French government contracted with SGN to construct a plutonium reprocessing facility in Taiwan. In 1975, France began to transfer

---

132. On the cancelled Argentina-Libya plutonium reprocessing deal, see Jones et al., *Tracking Nuclear Proliferation*, p. 224.

133. Ramon H. Myers and Jialin Zhang, *The Struggle across the Taiwan Strait* (Stanford, CA: Hoover Institution Press, 2006), p. 3.

134. The following account of the cancelled France-Taiwan plutonium reprocessing deal draws from Weissman and Krosney, *Islamic Bomb*, pp. 151–153; and Spector, *Nuclear Proliferation Today*, pp. 342–344.

some of the component parts for the plutonium reprocessing facility to Taipei. As in the Israeli case, when the United States learned of this sensitive nuclear transaction, it intervened and pressured both sides to cancel the deal. The United States threatened to cut off all economic and military aid to Taiwan. At this time, Taiwan was still protected by a U.S. defense pact and was unwilling to upset the United States over the sensitive nuclear imports. Taiwan notified France that it was cancelling the contract for the plutonium reprocessing facility. In short, Taipei chose the U.S. Seventh Fleet over the French reprocessing facility.

In 1976, the Ford administration followed through to ensure that Taiwan would not be able to use the parts that it did receive from France to develop a plutonium reprocessing facility. The United States forced Taiwan to dismantle the facility and return some of the components to the United States. In the words of one Taiwanese scientist, "After the Americans got through with us, we wouldn't have even been able to teach physics here on Taiwan."[135]

In sum, states that rely on a superpower security guarantee are vulnerable to superpower pressure and are less likely to participate in sensitive nuclear transactions. As we saw in the cases of Argentina and Libya in 1985 and France and Taiwan in 1975, when superpower-dependent states stumble into an arrangement to provide or receive sensitive nuclear assistance, it is unlikely that the transaction will actually be consummated. The superpower patron will intervene, and the dependent state will withdraw from the sensitive nuclear agreement. This discussion of additional cases lends support to the argument that France's superpower independence facilitated its sensitive nuclear cooperation with Israel.

The strategic theory of nuclear proliferation finds strong support in the French-Israeli case. France lacked the ability to project power over Israel, meaning that it could provide sensitive nuclear assistance to Israel without constraining its own military freedom of action. This supports Hypothesis 1. In support of Hypothesis 2, we saw that France was motivated to provide sensitive nuclear assistance to Israel in order to constrain their common rival, Egypt. And, consistent with Hypothesis 3, France was not dependent on the superpowers, allowing France to provide sensitive nuclear assistance to Israel, despite U.S. opposition.

Alternative Explanations

The strategic theory of nuclear proliferation provides a more persuasive explanation for French-Israeli nuclear cooperation than competing economic, institutional, or nuclear weapons possession arguments.

---

135. Unidentified Taiwanese scientist, as cited in Weissman and Krosney, *Islamic Bomb*, p. 153.

Economics

An explanation based on economic motivations fails to hold up under the weight of the evidence. French and U.S. approaches to nuclear proliferation differed starkly, but their economic positions were quite similar. Both countries were advanced industrial countries with rapidly growing economies. France and the United States both enjoyed growth rates of about 5 percent per year in this time period.[136] Neither state was driven to export sensitive nuclear technology by economic desperation or by poor economic performance. Nor was either state especially dependent on trade with Israel. In this time period, trade with Israel accounted for less than one tenth of one percent of total GDP for both the United States and France.

More nuanced economic explanations also fail. Nicolas Jabko and Steven Weber have argued that, because of its dependence on international trade, France may be more likely than other states to export nuclear technology. They explain that France's "national, large-scale, and autonomous nuclear complex...would generate significant economic pressures to export technology and facilities."[137] In other words, Jabko and Weber argue that France's powerful nuclear industry should be able to lobby successfully for the export of nuclear materials and facilities. Exports could help the nuclear industry to turn a profit in good times and to stay afloat when times are bad. If Jabko and Weber are correct, we should expect that the domestic nuclear complex has significant influence over nuclear export decisions and that the nuclear industry can gain substantial economic rewards for exporting sensitive nuclear materials and technology. But neither of these assumptions is correct.

The nuclear industry is rarely a significant player in decisions to export sensitive nuclear materials and technology. The French nuclear industry may have been looking to expand its market, but industry's influence over government policy was not decisive when it came to sensitive matters. As one expert of the French nuclear industry explains, "Industry's role in formulating foreign policy on export matters is only as large as the government allows it to be."[138] This statement is supported by the evidence in the French-Israeli case. In the available record, senior French officials repeatedly invoked strategic considerations when advocating nuclear transfers to Israel but never once so much as hinted at an economic motivation.

Second, there is not much money to be made by exporting sensitive nuclear materials and technology. The sums of money that a state can generally expect

---

136. All economic data are from Gleditsch, "Expanded Trade and GDP Data."
137. Jabko and Weber, "A Certain Idea," p. 126.
138. Bujon de L'Estang, Francois, "The Delicate Balance," in Jones et al., eds., *Nuclear Suppliers and Nonproliferation*, p. 139.

to gain from the provision of sensitive nuclear assistance are modest relative to the sums that can be earned from the export of conventional military technologies or from the sale of nuclear reactors and other civilian nuclear technologies. As *Les Echos,* the journal of the French AEC, explained in an article in the 1970s, "The sale of a reprocessing plant in itself is a pretty small affair on the scale of nuclear contracts; a few hundred million francs...where a power plant is figured at several billions."[139]

Moreover, France had already established lucrative contracts for military and civilian nuclear cooperation with Israel and would not have had substantial economic incentive to provide more sensitive nuclear technologies. In 1955, France secured a conventional military arms deal with Israel estimated to be worth over $100 million.[140] By September 1956, France had also secured a contract to export a small nuclear reactor to Israel, estimated to be worth about $80 million.[141] The final price tag on total French nuclear exports to Israel in this period, including the large, plutonium-producing reactor, the underground reprocessing facility, and the nuclear weapons-related assistance, is not known with precision. Perhaps the most reliable estimate comes from Shimon Peres, who claims that the total sum for the nuclear imports from France was about $80 million.[142] In other words, the total price tag that the French government and firms received for the entire nuclear package was roughly equivalent to the market rate for a single nuclear reactor. France's economic returns for providing sensitive nuclear assistance to Israel appear to have been marginal at best.

Institutions

Institutional and related normative explanations also fail to find support in this case. International institutions aimed at stemming nuclear proliferation had not yet been established. French-Israeli nuclear cooperation occurred before the NPT was opened for signature in 1968. Thus, neither the United States nor France was a member of the NPT, yet their behavior vis-à-vis Israeli's nuclear program was quite different.

An argument that focuses on the role of international norms also fails to explain the case. One could argue that in the pre-NPT era, a strong norm against nuclear proliferation had not yet developed, providing France with a permissive environment in which to provide sensitive nuclear assistance. Yet, this argument

---

139. As cited in Weissman and Krosney, *Islamic Bomb,* p. 165.
140. Peres, *Battling for Peace,* p. 120.
141. Cohen, *Israel and the Bomb,* pp. 52–55.
142. Peres, *Battling for Peace,* p. 119; and Pinkus, "Atomic Power to Israel's Rescue," p. 130.

is unable to explain why the United States, operating in a similar normative environment, took such a strong stance against nuclear proliferation. Moreover, the argument that a norm against nuclear proliferation had not yet developed is belied by the fact that French officials were clearly concerned about the possible international reactions to French-Israeli nuclear cooperation. Even President de Gaulle himself fretted over the likely negative international reaction to French-Israeli nuclear cooperation. French officials apparently believed that it was inappropriate for France to provide sensitive nuclear assistance to Israel, but they decided to do so anyway. This case casts serious doubts on claims that international norms against nuclear proliferation pose a significant constraint on state decisions to provide sensitive nuclear assistance.

Nuclear Weapons Possession

Strategic arguments that claim that nuclear weapon states have incentives to prevent nuclear proliferation in order to maintain the exclusivity of the nuclear club do not find support in the Israeli case. France and the United States were both nuclear weapon states, but, contrary to the nuclear exclusivity argument, only one of these states was opposed to nuclear proliferation in Israel. France became a nuclear weapon state in 1960 but was still willing to provide sensitive nuclear assistance to Israel until 1965. It appears, therefore, that France was not very concerned about keeping the hoi polloi out of the nuclear club. Moreover, the evidence indicates that U.S. officials opposed Israel's nuclearization due to the constraints it would have placed on America's conventional military freedom of action, not because they wanted to maintain a nuclear advantage.

Conclusion

The French-Israeli case provides clear support for the strategic theory of nuclear proliferation presented in this book. The U.S. ability to project military power over Israel meant that Israel's nuclearization would undermine America's strategic position in the region. Furthermore, the United States was not threatened by Israel's regional rivals and, therefore, could not constrain a common enemy by helping Israel acquire nuclear weapons. Moreover, because the United States was a superpower, it was threatened by nuclear proliferation anywhere in the international system. Thus, the United States refused to provide sensitive nuclear assistance to Israel, and it intervened to prevent French-Israeli nuclear cooperation. France's inability to project military power over Israel meant that France would not incur strategic costs if Israel acquired nuclear weapons. France, unlike the United States, could provide sensitive nuclear assistance to Israel without constraining its own military freedom of action. Moreover, France was able to provide sensitive nuclear assistance to Israel as part of a strategy to constrain Egypt, an enemy it shared with Israel. Furthermore, France did not depend on the United States to provide for its security, enabling it to continue with the sensitive nuclear transaction in the face of U.S. pressure to halt it.

The strategic theory of nuclear proliferation can also explain change over time in this important case. France halted its sensitive nuclear transfers to Israel in the early to mid-1960s for two primary reasons. First, as the Algerian War came to an end, France no longer had a strategic incentive to arm Israel in order to constrain Egypt. Second, and perhaps more important, France had already provided Israel with everything that it needed to produce nuclear weapons. Israel was well on its way to becoming a nuclear power and did not require additional assistance from Paris. On the other hand, Washington initially opposed nuclear proliferation to Israel because it felt that nuclear weapons in Tel Aviv would constrain its military freedom of action. In a move that the United States would repeat in many historical cases of nuclear proliferation, however, Washington was eventually willing to turn a blind eye to the existence of Israel's nuclear arsenal after it failed in its bid to stop nuclear proliferation. When Israel acquired nuclear weapons, Washington's calculus changed. U.S. officials decided that there was little chance of denuclearizing Israel. Recognizing a fait accompli, the United States decided to engage Israel as an ally against the Soviet Union and its encroachment in the region through its Arab proxies.

Additional shadow cases bolster the logic of the argument. An analysis of the Soviet Union's approach to Israel's nuclear program demonstrated that a superpower other than the United States also concluded that nuclear proliferation in Israel would constrain its conventional military freedom of action. So the Soviet Union, like the United States, took steps to stop Tel Aviv from acquiring the bomb. Brief studies of instances of cancelled sensitive nuclear cooperation agreements between Argentina and Libya in 1985 and France and Taiwan in 1975 demonstrated that states that are vulnerable to superpower pressure are less likely to participate in sensitive nuclear transfers. Argentina and Taiwan both enjoyed security guarantees from the United States, and they both cancelled proposed sensitive nuclear cooperation agreements under U.S. pressure.

This chapter has provided strong empirical support for the strategic theory of nuclear proliferation. But, perhaps the France-Israeli case is unique. Can the strategic theory of nuclear proliferation help us to explain other cases? Would a detailed analysis of a broader set of cases also provide support for the strategic theory of nuclear proliferation? We will find out in the next chapter.

4
COMMON ENEMIES, GROWLING DOGS, AND A. Q. KHAN'S PAKISTAN
Nuclear Supply in Other Countries

To this point, the book has provided quantitative evidence and in-depth case analysis that has provided strong support for the strategic theory of nuclear proliferation. Although the evidence presented to this point may be persuasive, a skeptical reader may glance at the list of cases of sensitive nuclear assistance listed in appendix C and wonder whether the theory can explain a broader set of cases. Is the "enemy of my enemy is my customer" logic unique to the France-Israel case, or has the desire to constrain a common enemy encouraged other states to export sensitive nuclear materials and technology? What about the cases that appear to contradict the theory? The Soviet Union was a superpower that enjoyed the ability to project military power over the entire international system, yet it provided sensitive nuclear assistance to neighboring China from 1958 to 1960. Why? Further, what about the dogs that did not bark? How do we explain the cases where countries should have provided sensitive nuclear assistance, but have not done so? India and Israel fall into this category. Finally, does the strategic theory of nuclear proliferation apply to the Cold War only? Or can it also illuminate recent cases, such as Pakistan's sensitive nuclear assistance, with the help of Pakistani scientist A. Q. Khan, to Iran, Libya, and North Korea? This chapter addresses these questions in a series of short case studies.

China to Pakistan, 1981–1986

From 1981 to 1986, the People's Republic of China (PRC) provided Pakistan with sensitive nuclear assistance.[1] China helped Pakistan construct and operate uranium enrichment facilities, transferred enough HEU for two nuclear weapons, and shared a nuclear bomb design with Pakistan. Why did China assist Pakistan's nuclear weapons program?

An analysis of this case provides strong support for the strategic theory of nuclear proliferation. China lacked the ability to project power over Pakistan, meaning that nuclear proliferation in Pakistan would not constrain China's military freedom of action. Further, China and Pakistan were threatened by a common enemy, India, motivating China to provide sensitive nuclear assistance to Pakistan in order to constrain a shared rival. Finally, China enjoyed the superpower dependence granted by a nuclear weapons arsenal and was able to continue the sensitive nuclear transactions despite superpower opposition.

China's Power Projection

China lacked the ability to project conventional military power over Pakistan. Although China enjoyed a large aggregate military advantage and had the ability to launch air strikes (including with nuclear weapons) against Pakistan, geographical barriers would have prevented China from fighting a full-scale conventional military ground war on Pakistani territory. This allowed China to provide sensitive nuclear assistance to Pakistan without greatly constraining its own military freedom of action.

Remember that the purpose of the military analysis is not to examine whether China would have wanted to invade Pakistan in this time period. In fact, China and Pakistan enjoyed friendly relations, and it is highly unlikely that China harbored any intentions of invading Pakistan. Yet, power-projecting states are threatened by nuclear proliferation even to states with which they enjoy friendly relations. Therefore, as in previous cases, I analyze China's ability to invade and fight a full-scale conventional war in Pakistan as a proxy measure of a key independent variable: China's ability to project power over Pakistan.

There is, of course, no question that China enjoyed a vast military superiority over Pakistan. China had a larger population, more men under arms, higher

---

1. In the 1990s, China also helped Pakistan with its reprocessing facility at Chasma and, in 1995, China exported five thousand ring magnets, component parts for uranium enrichment using the gaseous centrifuge method, to Pakistan. These transfers occurred after Pakistan is believed to have achieved a nuclear weapons capability and are not recorded as cases of sensitive nuclear assistance. See appendix C for more information.

levels of military spending, and a larger industrial base.[2] Yet, geographical barriers would have prevented China from fighting a full-scale conventional war on Pakistani territory. One method by which China could have attempted to invade Pakistan would have been by land across their common border.[3] But, the China-Pakistan border is shrouded in a particularly treacherous stretch of the Himalayan Mountains, and the terrain would have prevented China from launching a ground invasion. China could have conceivably sent small military units into Pakistan across the Karakoram Highway that links the two countries, but the highway was only completed in 1986, after the sensitive nuclear transfers were completed. Moreover, even with the highway in place, it is highly unlikely that China could have launched a full-scale ground invasion of Pakistan. The Karakoram Highway is narrow (ten meters) and incapable of supporting heavy vehicles (not to mention tanks), and the surrounding terrain is too difficult to allow for a full-scale military invasion in the face of a hostile defense.[4] Further, China lacks the advanced lift capabilities that would have allowed it to airlift forces over the mountains into Pakistan.[5]

Some may question whether the Sino-Indian War of 1962 demonstrates that China indeed possessed the ability to project power across the Himalayan Mountains.[6] On the contrary, a careful examination of the Sino-Indian War illustrates China's inability to project power across the Himalayas. During the Sino-Indian War, China did not invade and fight a full-scale conventional military ground war. Rather, the military engagements consisted of encounters between small military units of between ten and a couple of hundred soldiers. These engagements were not large, set-piece battles but rather small skirmishes over military observation posts. Furthermore, it is extremely unlikely that, had it wanted to, China could have escalated the Sino-Indian War and fought a full-scale conventional military conflict on Indian soil. The high elevation and rough terrain presented both the Chinese and Indian militaries with insurmountable logistical difficulties. Moreover, the Aksai Chin salt flats, where the Sino-Indian War was fought, provides

---

2. Data on military power are from Correlates of War Project, *National Material Capabilities Data Documentation*.

3. On China's use of military force to settle territorial disputes, see M. Taylor Fravel, "Power Shifts and Escalation: Explaining China's Use of Force in Territorial Disputes," *International Security*, Vol. 32, No. 3 (Winter 2007/2008), pp. 44–83.

4. On the Karakoram Highway and its impact on the region, see, for example, Hermann Kreutzmann, "The Karakoram Highway: The Impact of Road Construction on Mountain Societies," *Modern Asian Studies*, Vol. 25, No. 4 (October 1991), pp. 711–736.

5. On lift capacity and force projection see, for example, Dunnigan, *How to Make War*.

6. For a military analysis of the Sino-Indian War, see James Barnard Calvin, *The China-India Border War (1962)* (Marine Corps Command and Staff College, April 1984); and John W. Garver, *Sino-Indian Rivalry in the Twenty-First Century* (Seattle: University of Washington Press, 2001).

a far more welcoming terrain for military conflict than the much more challenging China-Pakistan border. The Aksai salt flats are at lower elevation and provide a relatively flat battle terrain. In contrast, the China-Pakistan border is at a higher elevation and is marked by steep and jagged mountain peaks. In sum, mountainous terrain would have prevented China from fighting a full-scale conventional military war in India in 1962, and a hypothetical military operation against Pakistan would have only been more difficult.

The only other means by which China could have conceivably projected power into Pakistan would have been through an amphibious invasion. But, China never developed an effective amphibious invasion capability. In fact, most military analysts conclude that, to this day, China lacks the military capability to launch an amphibious invasion of Taiwan, a territory a mere one hundred miles from China's own shore.[7] Clearly, an amphibious invasion of Pakistan, across thousands of miles of blue water, would have been unthinkable.

In sum, by providing sensitive nuclear assistance to Pakistan, China did not impose significant strategic constraints on itself. We lack evidence in the English language about how Chinese officials assessed the threat of nuclear proliferation in Pakistan. But given China's inability to project power into Pakistan, it is unlikely that Chinese officials worried that nuclear weapons in Pakistan would constrain China's conventional military freedom of action.

China's Common Enemies

China provided sensitive nuclear assistance to Pakistan in order to impose strategic costs on a shared enemy: India. China and India have been entangled in a decades-long state of enmity, fueled in part by an unsettled border dispute.[8] The territorial feud led to direct military clashes between China and India in 1962 (as discussed above), in 1967, and again in 1986 and 1987. For this and other reasons, China viewed India as a key regional rival.

Similarly, India and Pakistan are also engaged in an enduring rivalry.[9] Following Pakistan's independence in 1947, India and Pakistan fought in a series

---

7. Office of the Secretary of Defense, *2008 Annual Report on the Military Power of the People's Republic of China*.
8. On the China-India rivalry, see, for example, Garver, *Sino-Indian Rivalry*; Fravel, "Power Shifts and Escalation"; Alastair Lamb, *The Sino-Indian Border in Ladakh* (Canberra: Australian National University Press, 1973); Neville Maxwell, *India's China War* (New York: Pantheon Books, 1970); and Allen S. Whiting, *The Chinese Calculus of Deterrence* (Ann Arbor: University of Michigan Press, 1975).
9. On the India-Pakistan rivalry, see, for example, J. N. Dixit, *India-Pakistan in War and Peace* (New York: Taylor and Francis, 2007); and T. V. Paul, ed., *The India-Pakistan Conflict* (Cambridge, U.K.: Cambridge University Press, 2005).

of bloody wars. From 1947 to 1949, India and Pakistan clashed over control of Kashmir. Kashmir was the site of a second major war between the two countries in 1965. A civil war between East and West Pakistan, in 1971, led to an Indian military invasion of Pakistan. And Pakistan's nuclear weapons program was launched largely in response to the everpresent Indian threat.[10]

By providing sensitive nuclear assistance to Pakistan, China could impose strategic costs on a shared rival and confine India to regional power status. After all, China would have wanted to prevent India from rising to become a near-peer competitor in Asia. In particular, a nuclear weapons arsenal in Pakistan would complicate India's strategic calculations, refocus India's strategic attention away from Beijing and toward Islamabad, prevent India from dominating South Asia, and undermine Indian aspirations of challenging China in East Asia.

The strategic intent of China's sensitive nuclear assistance to Pakistan is evident in the content and context of the transfers. The nuclear transfers included materials that could only have been used in a military nuclear program, such as weapons-grade fissile material and nuclear weapon designs. Furthermore, the nuclear assistance occurred alongside ballistic missile and other conventional military transfers, demonstrating China's desire to strengthen Pakistan's military might.[11]

Scholars who have written on the Sino-Pak strategic relationship are unanimous in finding that China provided sensitive nuclear assistance to Pakistan in order to constrain India. For example, Alastair Iain Johnston has argued that China's motivation to provide arms to Pakistan was to "help divert Indian military resources away from China."[12] T. V. Paul maintains that Chinese nuclear transfers to Pakistan "derive largely from Chinese concerns about the regional balance of power and are part of a Chinese strategy to pursue containment in its enduring rivalry with India."[13] Thomas Reed and Danny Stillman write that there was a "determination in China to support India's enemies, whoever they may be. The principal beneficiary turned out to be Pakistan."[14] In sum, according to Gordon Corera, China provided sensitive nuclear assistance to Pakistan because it "was keen to see more nuclear powers in the world and particularly to see its rival India kept at bay."[15]

---

10. On the relationship between the nuclear programs in Pakistan and India, see, for example, Perkovich, *India's Nuclear Bomb*.
11. On other types of military related transfers between China and Pakistan, see, for instance, T. V. Paul, "Chinese-Pakistani Nuclear/Missile Ties and Balance of Power Politics," *Nonproliferation Review*, Vol. 10, No. 2 (Summer 2003), pp. 21–29.
12. Alastair Iain Johnston, "International Structures and Chinese Foreign Policy," in Samuel S. Kim, ed., *China and the World* (New York: Westview, 1998), p. 73.
13. Paul, "Chinese-Pakistani Nuclear/Missile Ties," p. 22.
14. Reed and Stillman, *Nuclear Express*, p. 158.
15. Corera, *Shopping for Bombs*, p. 45.

China's Superpower Dependence

The United States, a global superpower, intervened in an attempt to prevent Chinese-Pakistani nuclear cooperation. China was able to continue providing sensitive nuclear assistance, despite U.S. objections, because China enjoyed superpower independence.

Beginning in the early 1970s, the United States led an international effort to bring China into the international nonproliferation regime.[16] For decades the United States applied diplomatic pressure in a failed attempt to convince China to join the IAEA, the NPT, and the NSG. China was unwilling to join formal institutions that placed restrictions on the conditions under which it could export nuclear technologies. In the mid-1980s and early 1990s, U.S. officials were aware of, and concerned about, Chinese nuclear exports to Pakistan, Iran, and Algeria. In response, the United States implemented an approach that combined carrots and sticks to dissuade China from exporting sensitive nuclear material and technology. In the 1980s, the United States offered China civilian nuclear technology, conventional weapons systems, military technology, advanced computers, and telecommunications technology, in exchange for an agreement to meet U.S. nonproliferation standards. It may seem puzzling that the United States was willing to provide nuclear assistance to China, a potential rival state. These were, however, nonsensitive civilian nuclear transfers, which would not have directly strengthened China's nuclear weapons program. More important, however, they were part of an overall attempt at influencing China's nuclear nonproliferation behavior. U.S. attempts at influence continued with the 1985 Congressional Resolution of Approval of the U.S.-China Agreement for Nuclear Cooperation, which threatened to cut off U.S. nuclear-related exports to China if China failed to comply with U.S. nonproliferation requirements on its nuclear exports. In addition, the 1994 Nuclear Proliferation Prevention Act required a cutoff of U.S. financing to any country which "has willfully aided and abetted any non-nuclear weapon state" to acquire HEU or plutonium.[17]

U.S. pressure did not, however, prevent China from providing sensitive nuclear assistance to Pakistan. China was able to continue with the sensitive nuclear transactions in part because it enjoyed superpower independence. China had acquired a nuclear weapons arsenal in 1964, granting it security independence long before it initiated the sensitive nuclear transfers to Pakistan. Furthermore, China did not possess a formal defense pact with either the Soviet Union or the United States, meaning that China was not relying on a superpower to provide for its

---

16. On U.S. efforts to bring China into the nonproliferation regime, see, for example, Jones et al., *Tracking Nuclear Proliferation,* pp. 49–70; and Corera, *Shopping for Bombs.*
17. Nuclear Proliferation Prevention Act of 1994.

security. In fact, during the Cold War, China maintained enough independence from the superpowers that it was able to switch from one side to the other, and back again.

Because China did not depend on a superpower to provide security, it was able to continue the sensitive nuclear transfers, despite superpower pressure. China probably considered the international displeasure that would result from the provision of sensitive nuclear assistance, but the prospect of international condemnation was not enough to deter the sensitive nuclear transfers to Pakistan.

In recent years, China has come to exercise more restraint in its nuclear exports. China joined the IAEA in 1984, the NPT in 1992, and the NSG in 2004. China has also adopted a system of tougher export controls and has agreed to refrain from exporting nuclear materials and technology to nonnuclear weapon states unless the materials and technologies are placed under IAEA safeguards. It is doubtful, however, that this change in China's approach to nuclear proliferation was primarily the result of superpower pressure. As I will argue in the conclusion of this chapter, China's change of heart on nuclear export matters may be driven more by the rise of Chinese military power than by any other factor.

In sum, all three hypotheses of the strategic theory of nuclear proliferation find strong support in this case. Consistent with Hypothesis 1, China lacked the ability to project power over Pakistan, allowing it to provide sensitive nuclear assistance without constraining its own military freedom of action. China provided sensitive nuclear assistance in order to constrain a common enemy, validating Hypothesis 2. And, in support of Hypothesis 3, China, a state that was not dependent on a superpower, provided sensitive nuclear assistance despite superpower pressure.

Alternative Explanations

Alternative explanations that highlight the role of economics, international institutions, or nuclear weapons possession, cannot account for China's sensitive nuclear transfers to Pakistan.

ECONOMICS

There is little evidence to support an economic explanation for China's sensitive nuclear exports to Pakistan. Simple economic accounts fall short. China was not in dire economic straits when it provided sensitive nuclear assistance to Pakistan. On the contrary, in this period, China's economy was growing at a rapid rate. For example, in 1984, China's economy increased by an eye-popping 12 percent.[18]

---

18. All economic data are from Gleditsch, "Expanded Trade and GDP Data."

There is no support for the idea that China exported sensitive nuclear materials and technology because economic conditions forced its leaders to export anything in order to earn hard currency.

Neither is there support for more nuanced economic explanations. China was not particularly dependent on international trade with Pakistan, which accounted for only 0.02 percent of China's GDP in this period. This placed the Sino-Pak dyad in only the sixty-fifth percentile of most-trade-dependent dyads in the world from 1951 to 2000. Clearly, China did not help Pakistan because it was fearful of losing a critical trading partner.

INSTITUTIONS

Explanations that emphasize international institutions and international norms find mixed support in this case. China was not a member of the NPT or the NSG when it transferred sensitive nuclear materials and technology to Pakistan. This outcome is consistent with the expectation that countries that are not members of the international institutions of the nuclear nonproliferation regime will be more likely to provide sensitive nuclear materials and technology. There are, however, two obvious problems with an institutional explanation for this case. First, as with other institutional explanations, it is hard to know whether the observed relationship between international institutions and the observed outcome is the result of a constraining or screening effect. In other words, it is difficult to disentangle whether China provided sensitive nuclear assistance to Pakistan because it was not a member of the NSG or the NPT, or if it did not join the NSG and the NPT because it had an underlying propensity to export sensitive nuclear materials and technology. Given China's persistent reluctance to join these institutions at the same time that it was exporting sensitive nuclear materials and technology and China's later willingness to sign the NPT, after the transactions had ceased, it appears that the latter could have been the case. Second, the institutional explanation cannot explain important within-case variation in the China-Pakistan case. For example, why did China provide sensitive nuclear assistance to Pakistan but not other states? Why did China, but not India, another state that did not belong to either the NSG or the NPT, provide sensitive nuclear assistance to Pakistan? The institutional explanation cannot provide answers to these questions. It is only by incorporating the conditions identified by the strategic theory of nuclear proliferation that one can explain the details of this case. China provided sensitive nuclear assistance to Pakistan and not other states because Beijing and Islamabad shared a common rival. And India did not provide nuclear assistance to Pakistan because it did not want to undermine its own security environment.

Related normative explanations that contend that the establishment of the NPT created a norm that renders states reluctant to provide sensitive nuclear assistance find no support in this case. An argument about global nuclear proliferation norms cannot explain the clear within-case variation. China and the United States existed in the same normative milieu, but China actively aided Pakistan's nuclear program, while the United States opposed nuclear proliferation in Pakistan. While the United States probably could have done more to prevent nuclear proliferation in Pakistan, it always preferred that Pakistan remain nonnuclear and never considered aiding Pakistan in its quest to acquire nuclear weapons. The normative argument cannot account for this difference between U.S. and Chinese behavior. Moreover, if the normative argument is correct, the norm against nuclear proliferation should have been very strong by the mid-1980s because the NPT had been in existence for over a decade. Yet, China was perfectly willing to provide sensitive nuclear assistance to Pakistan at this time, suggesting either that a strong norm against nuclear proliferation did not exist or that even powerful nuclear nonproliferation norms did not pose a significant constraint on Chinese behavior.

NUCLEAR WEAPONS POSSESSION

Arguments that contend that nuclear weapon states have an incentive to prevent nuclear proliferation do not find support in this case. China exported sensitive nuclear materials and technologies to Pakistan at a time when it possessed nuclear weapons. Clearly, China was not overly concerned about keeping additional states out of the nuclear club.

USSR to China, 1958–1960

From 1958 to 1960 the Soviet Union provided the PRC with sensitive nuclear assistance.[19] The Soviet Union agreed to help China construct a plutonium reprocessing facility, a uranium enrichment plant, and to provide Beijing with a prototype nuclear weapon. In the end, Moscow did not fully deliver on its nuclear promises, but it did provide China with designs and key component parts for the Lanzhou uranium enrichment facility and the Jiuquan plutonium reprocessing plant. Why did Moscow provide sensitive nuclear assistance to Beijing?

The details of this case, at first glance, may appear to contradict the strategic theory of nuclear proliferation. The Soviet Union was a global superpower.

---

19. On nuclear and military cooperation between the Soviet Union and China, see John W. Lewis and Litai Xue, *China Builds the Bomb* (Stanford, CA: Stanford University Press, 1988); Holloway, *Stalin and the Bomb*; Sergei Goncharenko, "Sino-Soviet Military Cooperation," in Odd Arne Westad, ed., *Brothers in Arms* (Washington, D.C.: Woodrow Wilson Center Press, 1998); and Reed and Stillman, *Nuclear Express*, pp. 84–113.

By providing sensitive nuclear assistance to China, it was arming a state over which it had the ability to project conventional military power and risked constraining its own military freedom of action.

A careful examination of this case, however, reveals that the Soviet Union's nuclear assistance to China provides greater support for the strategic theory of nuclear proliferation than it does for rival explanations. The details of the case suggest that the Soviet Union, as a power-projecting state, was concerned about the strategic implications of nuclear weapons in China and was extremely reluctant to transfer sensitive nuclear materials and technology to a state on its border. When the Soviet Union eventually broke from character and provided sensitive nuclear assistance to neighboring China, it was motivated to constrain a common enemy, the United States. Furthermore, throughout this period, both the Soviet Union and the United States often behaved as prototypical superpowers, seeking to take measures to prevent nuclear proliferation in China.

The Soviet Union's Power Projection

The Soviet Union enjoyed a clear ability to project power over China. It maintained the largest and most powerful military on the Eurasian landmass, shared a common and unobstructed land border with China, and possessed amphibious invasion and airlift capabilities that would have allowed it to project power over much of the globe.[20] Thus, we would have expected the Soviet Union to have been threatened by the prospect of nuclear proliferation in China and to have been reluctant to provide China with sensitive nuclear assistance. While the Soviet Union did eventually provide sensitive nuclear assistance to China, a close examination of the case reveals that the Soviet Union behaved very much as the strategic theory of nuclear proliferation would have predicted. Moscow was initially unwilling to provide sensitive nuclear assistance to China, only provided China with outmoded sensitive nuclear technologies, refused to provide Beijing with assistance related to nuclear weapon design, and eventually reversed its decision and halted sensitive nuclear cooperation with Beijing altogether.

Sino-Soviet nuclear cooperation began with basic scientific exchanges and industrial development in 1955.[21] The Soviet Union provided China with a research reactor and brought Chinese scientists to Moscow to study nuclear physics.

---

20. On the Soviet Union's power-projection capability, see, for example, Dunnigan, *How to Make War*.

21. On Sino-Soviet nuclear cooperation, see Lewis and Xue, *China Builds the Bomb;* Holloway, *Stalin and the Bomb;* and Goncharenko, "Sino-Soviet Military Cooperation."

Beijing requested more sensitive forms of nuclear assistance from Moscow, but the Soviet Union was initially unwilling to provide China with sensitive nuclear materials and technologies. A continuing stream of refusals from Moscow led Chinese officials to conclude that the prospect of receiving sensitive nuclear assistance from the Soviet Union was hopeless. As late as September 1956, Chinese Vice-Premier Li Fuchun and other Chinese officials had definitively concluded that Moscow was unwilling to help China develop sophisticated weapons.

When the Soviets reversed course and began to transfer sensitive nuclear components to the Chinese in 1958, they were only willing to transfer outmoded technologies.[22] Chinese officials continually complained that the component parts for uranium enrichment and plutonium reprocessing that Moscow transferred were not the same parts that the Soviet Union used in its own, more modern facilities. Instead, the transfers consisted mostly of used parts left over from earlier Soviet efforts. Despite China's complaints, Soviet officials refused to transfer the latest enrichment and reprocessing technologies.

Moreover, Soviet officials were unwilling to share technology related directly to the design and construction of nuclear weapons. Much to the consternation of the Chinese, Moscow never followed through on its promise to provide China with a prototype atomic bomb, blueprints, and technical data related to nuclear weapons. The reluctance of the Soviet Union to share its most advanced sensitive nuclear technology with China was a constant source of friction between the two countries and is sometimes thought to be a key contributing cause to the Sino-Soviet split in 1960.[23]

Finally, by 1960, after less than a year and a half of nuclear assistance, Moscow decided to cease nuclear cooperation with China altogether.[24] Moscow cancelled the remaining nuclear transfers to Beijing. The Kremlin ordered Soviet scientists working on the Chinese nuclear project to return to the Soviet Union. Furthermore, as they returned, Soviet scientists were instructed to disassemble Chinese nuclear facilities and to repatriate nuclear materials and component parts back to the Soviet Union.

While we lack details about the Soviet Union's assessment about the threat of nuclear proliferation in China, the behavior of the Soviet Union indicates that Moscow was concerned about the threat posed by nuclear proliferation in China. The details of this case and our knowledge of the Soviet Union's threat assessment in similar cases strongly suggests that Moscow was concerned that nuclear proliferation in China would impose strategic costs on the Soviet Union.

---

22. Lewis and Xue, *China Builds the Bomb*.
23. Ibid.
24. Ibid.

The Soviet Union's Common Enemies

Despite its general opposition to nuclear weapons in China, the Soviet Union briefly reversed its position and supplied China with sensitive nuclear assistance from 1958 to 1960. We lack detailed information about the Soviet decision to aid China's nuclear ambitions, but the available evidence suggests that Sino-Soviet nuclear cooperation was driven largely by the desire to constrain the United States, an enemy shared by China and the Soviet Union.

According to the definition of rivalry introduced in chapter 2, both the Soviet Union and China were engaged in a rivalry with the United States. In the 1940s and 1950s, the U.S. military presence in East Asia threatened Chinese territorial integrity. At key moments during World War II, the Korean War, and at the height of the Taiwan Straits Crises of 1954–1955 and 1958, the United States and China engaged in militarized disputes.[25] By the late 1950s, the United States had also emerged as the Soviet Union's key Cold War adversary.[26] Moscow was concerned about the correlation of forces between capitalist and Communist systems worldwide and was threatened by the U.S. position in East Asia in particular. For example, during the Second Taiwan Straits Crisis, both Moscow and Beijing believed that a U.S. military attack on China could be imminent.[27] In a conversation with Soviet Foreign Minister Andrei Gromyko, Mao Zedong himself concluded that the United States could attack mainland China. He further proposed to Gromyko that if the United States penetrated deep into mainland China, the Soviet Union should retaliate and hit the United States "with everything you've got." The Soviet Union was concerned about the prospect of an escalation of the crisis into a superpower war. Considering the possibility of a superpower conflict over China, Gromyko replied that such an event "would not meet with a positive response by us. I can say that definitely."[28]

---

25. On U.S.-China relations during the Cold War, see, for example, Shu Zhang, *Economic Cold War* (Stanford, CA: Stanford University Press, 2002); Yafeng Xia, *Negotiating with the Enemy* (Bloomington: Indiana University Press, 2006); and John Lewis Gaddis, *We Now Know* (New York: Oxford University Press, 1998).

26. On U.S.-Soviet relations during the Cold War, see, for example, Ronald E. Powaski, *The Cold War* (New York: Oxford University Press, 1997); Gaddis, *We Now Know;* Odd Arne Westad, *The Global Cold War* (Cambridge, U.K.: Cambridge University Press, 2007); and Raymond Garthoff, *Détente and Confrontation* (Washington, D.C.: Brookings Institution, 1994).

27. On the Taiwan Straits Crises, see, for example, Robert Accinelli, *Crisis and Commitment: United States Policy toward Taiwan* (Chapel Hill: University of North Carolina Press, 1996); Michael Share, *Where Empires Collide* (Hong Kong: Chinese University Press, 2007); and Sukhumbhand Paribatra, *The Taiwan Straits Crisis of 1958* (Bangkok: Institute of Asian Studies, Chulalongkorn University, 1981).

28. Andrei A. Gromyko, *Memoirs* (New York: Doubleday, 1989), pp. 251–252.

The provision of nuclear assistance to China from 1958 to 1960 appears to have been part of a Soviet strategy to prevent direct U.S. military action against China and to contain U.S. power in East Asia. In this way, China as a junior ally could augment the Soviet Union's power vis-à-vis the United States. The Soviet Union's overarching goal of strengthening Chinese military power to balance the United States is evident in the massive arms transfers from the Soviet Union to China in this period. From 1955 to 1961, the Soviet Union transferred tanks, artillery, aircraft, and other conventional military hardware to Beijing.[29] The concurrence of the sensitive nuclear transfers with these large-scale conventional arms exports suggests that the nuclear materials and technology may have been part and parcel of a strategy to enhance China's defensive and deterrent capabilities.

The timing of the nuclear transfers further suggests that they were aimed at constraining U.S. military power in general and were also a direct response to the threat of U.S. invasion of China during the Second Taiwan Straits Crisis. Although China and the Soviet Union signed a nuclear cooperation agreement in 1955, the Soviet Union did not actually begin transferring sensitive nuclear materials and technology to China until 1958, nearly three years later. Furthermore, the Soviet Union officially cancelled the sensitive nuclear transactions in July 1959, shortly after the Second Taiwan Straits Crisis ended. The official cancellation date of the transfers suggests that once the imminent threat of U.S. intervention in the Taiwan Straits Crisis had passed, Moscow was no longer eager to see Beijing in possession of a nuclear arsenal.

Regional experts concur that Sino-Soviet military and nuclear cooperation was designed to constrain U.S. military power in East Asia. In the landmark book on China's nuclear program, John Lewis and Xue Litai wrote, "The convergence and articulation of security interests between the two Communist powers in turn profoundly influenced the Kremlin's decisions to support the Chinese nuclear program." Lewis and Litai continued, "The primary purpose [of China's nuclear program] was to strengthen the nation's defenses to meet a serious security threat, and the assistance provided by the Soviet Union in the initial phases of the nuclear program strongly suggests that the target was the United States."[30] In short, according to Sergei Goncharenko, Sino-Soviet military and nuclear cooperation took place because "to both Moscow and Beijing, the United States remained a strategic enemy."[31]

---

29. On Soviet military aid to China, see, for example, Westad, *Brothers in Arms.*
30. Lewis and Litai, *China Builds the Bomb*, pp. 41, 2.
31. Goncharenko, "Sino-Soviet Military Cooperation," p. 152.

The Soviet Union's Superpower Dependence

The Soviet Union was a superpower with global force-projection capabilities. Thus, it was threatened by nuclear proliferation in China and was willing to contemplate taking action, including the use of military force, to prevent Beijing from acquiring nuclear weapons.

The Soviet Union was a superpower that stood to inherit a more threatening strategic environment if China acquired nuclear weapons. Although for a very brief period in the late 1950s, the Soviet Union apparently saw the nuclearization of China as in its best interests, by the late 1960s, the Soviet Union reversed its position and opposed nuclear proliferation in China.[32] As the strategic theory of nuclear proliferation would expect from a superpower, in the late 1960s, the Soviet Union was willing to consider tough measures to denuclearize China. Indeed, in 1969, as the Soviet Union and China squared off in a border dispute, the Soviet Union threatened a military attack against China's fledgling nuclear arsenal.[33] The Soviet military made preparations for a preventive attack on China. In June of that year, Soviet bombers conducted mock attacks on targets in Siberia and Mongolia built to the specifications of China's nuclear facilities.[34] At the same time, Soviet diplomats prepared the world for the possibility of a preventive strike on China. On August 18, a Soviet diplomat approached a U.S. State Department official to inquire about what the U.S. reaction would be in response to a Soviet military strike against China's nuclear program.[35] On September 16, Victor Louis, a Soviet journalist with ties to the KGB wrote in the London *Evening News* that a military attack on China's nuclear facilities could be imminent. In the end, Moscow decided not to attack China's nuclear program, but their preparations demonstrate that Moscow was threatened by China's nuclear program and was willing to consider extreme measures to denuclearize China.

SHADOW CASE: U.S. SUPERPOWER DEPENDENCE

An examination of the U.S. response to nuclear proliferation in China provides further support for the idea that superpowers are particularly opposed to nuclear proliferation and act to prevent it. The United States, a global superpower, was worried about nuclear proliferation in China and took a number of measures to stop it. As we saw in chapter 1, U.S. officials concluded that nuclear proliferation

---

32. Burr and Richelson, "Whether to Strangle," p. 97.
33. On the Soviet Union's preparations for a preventive military strike on China's nuclear facilities, see Garthoff, *Détente and Confrontation*, pp. 236–237.
34. Dimitri K. Simes, "Soviet Policy toward the United States," in Joseph S. Nye, Jr., ed., *The Making of America's Soviet Policy* (New Haven: Yale University Press, 1984), p. 305.
35. Henry A. Kissinger, *The White House Years* (New York: Little, Brown, 1979), pp. 183–185.

would undermine America's strategic position in East Asia. For example, a June 1961 Joint Chiefs of Staff report predicted that China's "attainment of a nuclear capability will have a marked impact on the security posture of the United States and the Free World, particularly in Asia."[36] To prevent China from acquiring nuclear weapons, the United States considered a number of measures to destroy China's nuclear program, including the use of preventive military strikes.[37] U.S. officials seriously examined the military option. In 1963, the State Department's director of policy planning, Walt Rostow, ordered his staff to evaluate the pros and cons of a military strike on China's nuclear facilities. Rostow asked the staff to consider four options: U.S. air strikes, ground attacks by U.S. special operations forces, air strikes by Taiwan, and sabotage by agents of Taiwan. The U.S. military also prepared operational plans for a military strike. In the fall of 1963, the United States flew a U-2 spy aircraft over a nuclear reactor in China to assess the feasibility of a military strike. By late December, the JCS had developed an operational plan for a multiple-sortie air attack on China's nuclear facilities.[38]

The United States even attempted to elicit the Soviet Union's cooperation in a joint U.S.-Soviet military operation to destroy the Chinese nuclear program.[39] In a meeting with his advisors in 1963, President Kennedy wondered aloud: "You know, it wouldn't be too hard if we could somehow get kind of an anonymous airplane to go over there, take out the Chinese facilities—they've only got a couple—and maybe we could do it, or maybe the Soviet Union could do it, rather than face the threat of a China with nuclear weapons."[40] President Kennedy followed up on this idea by making a formal proposal to the Soviet Union. On July 14, 1963, President Kennedy sent U.S. official Averell Harriman to Moscow with instructions to sound out Soviet Premier Nikita Khrushchev's views on "limiting or preventing Chinese nuclear development and his willingness either to take Soviet action or to accept U.S. action aimed in this direction."[41] The Soviet Union did not respond positively in this instance, but the possibility of joint

---

36. Memorandum from the Joint Chiefs of Staff, "A Strategic Analysis of the Impact of the Acquisition by Communist China of a Nuclear Capability," June 26, 1961, as cited in Richelson, *Spying on the Bomb*, p. 144.

37. On U.S. considerations to use military force to stop China's nuclear program, see, for example, William Burr and Jeffrey T. Richelson, "A Chinese Puzzle," *Bulletin of the Atomic Scientists*, Vol. 53, No. 4 (July/August 1997), pp. 42–47; Burr and Richelson, "Whether to Strangle"; and Richelson, *Spying on the Bomb*, pp. 137–194.

38. See Richelson, *Spying on the Bomb*, pp. 153–156.

39. On U.S. efforts to elicit Soviet cooperation in a joint U.S.-Soviet strike on China's nuclear facilities, see Burr and Richelson, "A Chinese Puzzle," pp. 42–47; Burr and Richelson, "Whether to Strangle"; and Richelson, *Spying on the Bomb*, pp. 137–194.

40. John F. Kennedy, as quoted in Burr and Richelson, "Whether to Strangle," p. 54.

41. John F. Kennedy, as cited in Burr and Richelson, "A Chinese Puzzle," pp. 42–47.

U.S.-Soviet military action against China's nuclear program would continue to be considered at the highest levels of the U.S. government until China's nuclear test in 1964. On September 15, 1963, senior State Department officials again met to review possible steps to prevent China from becoming a nuclear power and decided to approach the Soviet Union a second time about a possible agreement to cooperate in preventive military action.[42] With the approval of President Johnson, Secretary of State Dean Rusk twice met with Soviet Ambassador to the United States Anatoly Dobrynin within two weeks to discuss the possibility of a joint U.S.-Soviet military action against China's nuclear program. The Soviet Union had ceased its nuclear cooperation with China more than three years earlier, but it was not yet ready to consider the use of force against China's nuclear program. Had the Soviet Union been more receptive to the U.S. proposals, it is possible that the two archenemies of the Cold War could have fought together on the same side against the common threat of nuclear proliferation in China.

In sum, the Sino-Soviet case provides some support for the strategic theory of nuclear proliferation. Hypotheses 2 and 3 are strongly supported by the evidence. Consistent with Hypothesis 2, the Soviet Union provided sensitive nuclear assistance to a state with which it shared a common enemy. And, as expected by Hypothesis 3, the world's two superpowers, the United States and the Soviet Union, intervened in an attempt to prevent nuclear proliferation in China. Only Hypothesis 1 finds less than unqualified support in this case. The outcome of Sino-Soviet cooperation is inconsistent with the expectations of Hypothesis 1 because the Soviet Union provided sensitive nuclear assistance to a state over which it enjoyed a power-projection capability. Nevertheless, a within-case examination provides some support for the expectations of the theory. The Soviet Union, a power-projecting state, was highly reluctant to provide China with sensitive nuclear assistance.

Alternative Explanations

While the strategic theory of nuclear proliferation cannot explain every contour of Sino-Soviet nuclear cooperation, it provides us with a much firmer basis for understanding than alternative theoretical perspectives.

ECONOMICS

Economic explanations cannot account for the nuclear transfers. In fact, sensitive nuclear exports to China presented economic burdens, not economic boons, to Moscow. The Soviet Union spent a colossal sum to help China with its nuclear

---

42. Ibid., p. 46.

program. Over the course of Sino-Soviet cooperation, Moscow financed Beijing's nuclear, military, and industrial acquisitions to the tune of over 100 billion rubles. According to some estimates, 100 billion rubbles was roughly 7 percent of the USSR's annual national income in this period.[43] Soviet leaders expected no reimbursement or economic gains from these transfers. Instead, they worried that they could no longer afford such largesse. According to Sergei Goncharenko, Soviet officials asked, "How much could Moscow provide in assistance [to China] without harming its own economic development?"[44] In short, Moscow aided China's nuclear program despite its economic interests, not because of them.

INSTITUTIONS

An institutional explanation receives little support in this case. Sino-Soviet nuclear cooperation occurred before the establishment of the nuclear nonproliferation regime. The absence of these international institutions may have provided the Soviet Union with an environment conducive to the international transfer of nuclear materials and technology. Yet, an institutional explanation cannot explain why the United States consistently opposed nuclear proliferation to China even though the nuclear nonproliferation regime had not yet been created. Nor can it explain why the Soviet Union changed tack and eventually opposed nuclear proliferation in China.

NUCLEAR WEAPONS POSSESSION

The Soviet Union possessed nuclear weapons at the time that it was actively exporting sensitive nuclear materials and technology to China. This casts further doubt on the argument that nuclear weapon states have strong incentives to limit the size of the nuclear club and are less likely to provide sensitive nuclear assistance.

IDEOLOGY

Though not one of the alternative explanations presented in chapter 1, one could plausibly argue that the Soviet Union helped China with its nuclear weapons program because the two countries shared a common Communist ideology. Yet, upon closer examination, we find that ideological affinity does not provide a compelling explanation for the Sino-Soviet case. Certainly, it is possible that the common Communist ideology of the USSR and the PRC may have played some role in facilitating cooperation, but this explanation raises more questions than it answers. An ideological explanation cannot account for why the Soviet

---

43. Goncharenko, "Sino-Soviet Military Cooperation," p. 160.
44. Ibid, p. 148.

Union provided sensitive nuclear assistance to China but was unwilling to provide sensitive nuclear assistance to any of the dozens of other states in the Communist bloc. If the Soviet Union was willing to help China because it was a fellow Communist state, we should have also expected it to provide sensitive nuclear assistance to other Communist countries like Poland, Yugoslavia, and North Vietnam. But Moscow did not help these countries. In fact, the Soviet Union steadfastly refrained from transferring sensitive nuclear technology to any other states in the Communist bloc.

Furthermore, if the glue of Communist ideology is what held Sino-Soviet nuclear cooperation together, we would have expected the sensitive nuclear transfers to occur much earlier. The ideological affinity between the Soviet Union and China was at its peak in the early and mid-1950s.[45] Yet, during this period, China was continually rebuffed by the Soviet Union in its repeated requests for sensitive nuclear assistance. By the mid-1950s, the Soviet Union and China began to part ways ideologically. Following the death of Joseph Stalin in 1953, Nikita Khrushchev assumed the position of premier of the Soviet Union. Mao never developed the close relationship with Khrushchev that he enjoyed with Stalin. Khrushchev and Mao continually tangled over who was the rightful leader of the world Communist movement, how best to promote the international spread of Communism, and whether and how to confront the U.S. threat, among other key ideological issues. In the words of John Lewis Gaddis, "It was a schism, in which true believers fell into a long and debilitating quarrel over what within their common faith was true and therefore to be believed.... Mao Zedong's dissent was for international communism, what the Protestant Reformation had been for the Roman Catholic Church."[46] The Sino-Soviet alliance could not last long under the pressure of this internal fracture; by the late 1950s and early 1960s the Sino-Soviet split had occurred. Yet, it was at the height of the ideological rift within Communism, on the eve of the Sino-Soviet split, when Moscow finally provided Beijing with sensitive nuclear assistance. In short, Moscow refused to provide China with sensitive nuclear assistance when the two countries shared a common Communist ideology—but agreed to do so when they did not. These facts cast serious doubt on an ideological explanation for Sino-Soviet nuclear cooperation.

The Dogs That Did Not Bark

At this point the reader may wonder, if non-power-projecting states can provide sensitive nuclear assistance to constrain their enemies, why don't they do it

---

45. See, for example, Gaddis, "We Now Know."
46. Ibid, p. 212.

more often? How do we explain the countries that could have provided sensitive nuclear assistance, but did not? In other words, what about the dogs that did not bark?

This section addresses these questions in two parts. First, a careful reconsideration of the strategic theory of nuclear proliferation demonstrates that the set of conditions under which we should expect sensitive nuclear assistance to occur are rarer than they might initially appear. Keeping this set of scope conditions in mind, it is not surprising that sensitive nuclear assistance is a rare event. In fact, there are very few cases in which the strategic theory of nuclear proliferation would have predicted that a country should have provided sensitive nuclear assistance, but it failed to do so. There are, however, some obvious cases of dogs that did not bark. For this reason, the second part will examine two countries, Israel and India, that might have found it in their interests to provide sensitive nuclear assistance but did not.

Examining the Frequency of Sensitive Nuclear Assistance

A reconsideration of the strategic theory of nuclear proliferation demonstrates why sensitive nuclear assistance is a rare event. There are many conditions that combine to increase the risk of sensitive nuclear assistance, but these conditions are not often met. The following reexamination of each of these factors helps us to recast our expectations about the frequency with which sensitive nuclear assistance should occur. First, there are only nineteen states that are capable nuclear suppliers. There are, therefore, only a small number of states that have the opportunity to provide sensitive nuclear assistance. Second, many of the capable nuclear suppliers also have the ability to project power over potential nuclear recipients and, according to Hypothesis 1, refrain from providing sensitive nuclear assistance. Superpowers, states with the ability to project power over every other state in the international system, will be unlikely to provide sensitive nuclear assistance anywhere for fear of constraining their own military freedom of action. The list of capable nuclear suppliers also includes, however, a number of regional powers. Although these states lack global force-projection capabilities, they do possess the ability to project power over some neighboring states. Therefore, the strategic theory would also predict that these regional powers will be unlikely to provide sensitive nuclear assistance to those states over which they have the ability to project military power. Third, Hypothesis 2 posits that states provide sensitive nuclear assistance to constrain rival states, but there are a finite number of situations in which a potential supplier shares a common enemy with a potential recipient. Fourth, many capable nuclear suppliers are vulnerable to superpower pressure and, according to Hypothesis 3, do not provide sensitive nuclear assistance in order to avoid antagonizing a powerful patron. With these conditions in mind, it is possible to identify a few cases in which the theory predicts sensitive nuclear assistance, but sensitive nuclear assistance does not occur.

The Dogs That Growled but Did Not Bark

Israel and India are two states that, according to the strategic theory of nuclear proliferation, are at an elevated risk of providing sensitive nuclear assistance. Both countries possess the ability to supply nuclear materials and technology, have circumscribed force-projection capabilities, are threatened by powerful enemies, and enjoy the superpower independence afforded by nuclear weapons arsenals. Yet, neither state has provided sensitive nuclear assistance abroad. Why? A closer examination reveals that the forces identified by the strategic theory of nuclear proliferation have been in effect in these cases. While these countries have not yet provided sensitive nuclear assistance, Israel has come close and India may yet do so.

ISRAEL

Israel is a state that is at risk of providing sensitive nuclear assistance. Israel's ability to project military power is limited to its own region. Since its founding in 1948, Israel has been threatened by rival states that it could seek to constrain through the provision of sensitive nuclear assistance to other states. And, since 1967, it is widely believed to have possessed a nuclear weapons arsenal, granting it superpower independence.

Of course, it could be argued that Israel is not truly superpower-independent because, since 1967, Israel has relied heavily on U.S. political and military support. If, due to its relationship with the United States, Israel is best characterized as superpower-dependent, its restraint in the field of nuclear exports is easily explained by the strategic theory of nuclear proliferation; Israel refrained from providing sensitive nuclear assistance because it did not want to jeopardize its relationship with its key patron. There is a reasonable argument to be made, however, that Israel is best categorized as superpower-independent. It is undoubtedly true that the United States and Israel have enjoyed a close strategic relationship for decades, but the implications of this for coding Israel's superpower dependence are unclear. It could reasonably be argued that because of the strong pro-Israeli lobby in the United States and the perceived strategic importance of the Israel alliance to U.S. interests in the Middle East, the United States is at least as dependent on Israel as Israel is on the United States.[47] Furthermore, the coding rule explicated above clearly indicates that countries that possess a nuclear

---

47. On the pro-Israel lobby in the United States, see Mearsheimer and Walt, *Israel Lobby and U.S. Foreign Policy*.

weapons arsenal cannot be superpower-dependent. By this definition, Israel qualifies as superpower-independent, and its restraint in the field of nuclear exports once again seems puzzling.

According to the strategic theory of nuclear proliferation, therefore, Israel could potentially benefit by providing sensitive nuclear assistance to another state. It could provide sensitive nuclear assistance beyond its own sphere of influence without constraining its own military freedom of action. It could aid the nuclear program of a country with which it shared a common enemy in order to impose strategic costs on a rival state. And, since it enjoys the security independence provided by a nuclear arsenal, it would not have to worry much about antagonizing a superpower patron.

Despite the potentially favorable strategic conditions identified above, however, Israel has not provided sensitive nuclear assistance. While Israel did not provide sensitive nuclear assistance as I define it, it has come very close. In the late 1970s, Israel engaged in strategic cooperation with South Africa. Israel provided South Africa with ballistic missile technology.[48] It also supplied South Africa with tritium in 1977. Tritium is a nuclear material that can be used to increase the yield of a standard fission-type nuclear weapon to create a "boosted" nuclear bomb. Tritium does not qualify as sensitive nuclear assistance as I define it because sensitive nuclear assistance is that which helps a nonnuclear weapon state acquire nuclear weapons. While tritium can boost the yield of certain types of nuclear weapons, it cannot help a nonnuclear weapon state develop nuclear weapons. Israel may have also provided other types of nuclear assistance to South Africa. Although the idea has been dismissed by leading nuclear physicists, there is much speculation that Israel and South Africa may have conducted a joint nuclear weapons test off the coast of South Africa in 1979.[49]

Of course, determining whether or not Israel provided sensitive nuclear assistance in this case depends heavily on how sensitive nuclear assistance is defined. If I adopt an expanded definition of sensitive nuclear assistance that includes tritium, Israel did provide sensitive nuclear assistance to South Africa. In this case, Israel's sensitive nuclear transfers to South Africa are perfectly consistent with the expectations of the strategic theory of nuclear proliferation.

---

48. For the best source on Israeli-South African strategic cooperation, see Peter Liberman, "Israel and the South African Bomb," *Nonproliferation Review*, Vol. 11, No. 2 (Summer 2004), pp. 46–80.

49. For the official White House report provided by leading nuclear physicists that concluded that a nuclear test did not take place, see Jack Ruina, Louis Alvarez, William Donn, Richard Garwin, Riccardo Giacconi, Richard Muller, Wolfgang Panofsky, Allen Peterson, and Williams Sarles, "Ad hoc Panel Report on the September 22 Event," Executive Office of the President, Office of Science and Technology Policy, July 17, 1980. For an in-depth analysis of this assessment and subsequent analysis on the topic, see Richelson, *Spying on the Bomb*.

Even if we stick with the narrow definition and exclude Israeli–South African nuclear cooperation as an instance of sensitive nuclear assistance, however, the case of Israel still lends some support to the central argument of this book. Although Israel did not provide sensitive nuclear assistance as the strategic theory of nuclear proliferation would have predicted, it did come very close.

Furthermore, it is also plausible that an "enemy of my enemy is my customer" motivation was partly responsible for Israeli–South African strategic cooperation. South Africa's nuclear weapons program appears to have been developed in part to counter Soviet threats to South Africa in the late 1970s.[50] Soviet-backed Cuban troops in neighboring Angola supported a regime that was hostile to Pretoria. Mozambique, a country on South Africa's northeastern border, was also ruled by a Marxist-Leninist regime. South African officials viewed the Soviet Union's support of hostile regimes in neighboring states as a threat. South Africa's official nuclear doctrine envisioned nuclear weapons serving as a deterrent to an attack by Soviet-backed troops in neighboring states.

The common Soviet threat facing both Israel and South Africa may have been one factor that encouraged Israel to provide nuclear assistance to South Africa. Speaking about an exchange with South African military officials in 1977, Israeli Major General Avraham Tamir explained: "In the same way that we would describe the existential dangers facing Israel due to the Soviet Union's involvement and controls over Arab states, they also talked about the Soviet footholds around South Africa. Their conclusion was identical to ours: to defend against these dangers, it was necessary to develop [a]...nuclear capability. Obviously we had deep cooperation with them in all spheres."[51]

In sum, the conditions that facilitate sensitive nuclear assistance in other cases may have also encouraged Israel to engage in strategic cooperation with South Africa that came very close to sensitive nuclear assistance by my definition. While Israel did not quite make the decision to provide sensitive nuclear assistance to South Africa, it teetered perilously close to the edge.

INDIA

India is another state that has been at risk of providing sensitive nuclear assistance. India's force-projection capability is, at present, limited to South Asia; it is threatened by rival states, such as China, and could conceivably provide sensitive nuclear assistance to one of China's enemies in order to impose strategic costs

---

50. For the best sources on South Africa's nuclear program, see Peter Liberman, "Rise and Fall," pp. 45–86; Liberman, "Israel and the South African Bomb"; Mitchell Reiss, *Without the Bomb* (New York: Columbia University Press, 1988); and Mitchell Reiss, *Bridled Ambition* (Princeton, NJ: Princeton University Press, 1995).

51. Avraham Tamir, as cited in Liberman, "Israel and the South African Bomb," p. 55.

on Beijing; and it enjoys independence from superpower pressure afforded by a nuclear weapons arsenal. Yet India has never exported sensitive nuclear materials and technology to another state.

While India has not yet been compelled by its strategic conditions to provide sensitive nuclear assistance, there are signs that the strategic benefits that India could gain by exporting sensitive nuclear materials and technology are well understood in policymaking circles in New Delhi. Bahrat Karnad, a professor of national security studies at the Centre for Policy Research in New Delhi, argues that India should provide sensitive nuclear assistance to Vietnam and Taiwan to constrain China's conventional military freedom of action. In a recent book, Karnad writes, "India should, likewise, create precisely the kind of dilemmas for China that Beijing has created for it with respect to a nuclear weapons and missile-equipped Pakistan by arming Vietnam with strategic weapons" and by "cooperating with Taiwan in the nuclear and missile fields."[52] Karnad notes that a similar strategy worked for China. He writes that "China's masterful manipulation of the level of support to Pakistan as a means of influencing Indian policy and diluting New Delhi's perception of China as a threat has worked, but only because New Delhi seems to lack the nerve to impose heavy strategic costs on China by linking up with Vietnam and Taiwan...in a tit-for-tat policy."[53] While Karnad's views may not be representative of India's foreign policy elite, they demonstrate the pervasiveness of this type of thinking in countries positioned in security environments conducive to sensitive nuclear assistance. While India has not yet provided sensitive nuclear assistance, it may very well do so in the future.

There are also reasons to believe, however, that the probability that India will provide sensitive nuclear assistance may decline as time passes. India is a rising power.[54] It is investing heavily in military capabilities, including a blue-water navy, that could greatly increase New Delhi's ability to project power in the Indian Ocean and beyond. As India's power-projection capabilities increase, India will likely become more threatened by the proliferation of nuclear weapons in distant regions. In that event, New Delhi will have an important strategic reason to refrain from providing sensitive nuclear assistance. If India's military power continues to grow, there may be reason to believe that Professor Karnad's pleas will fall on deaf ears and New Delhi will maintain its unblemished record on

---

52. Bharat Karnad, *Nuclear Weapons and Indian Security* (New Delhi: Macmillan India, 2002), pp. xvii–xviii.
53. Ibid., p. xviii.
54. On India's rising power, see for example, Baldav Raj Nayar and T. V. Paul, *India in the World Order* (Cambridge, U.K.: Cambridge University Press, 2002); and C. Raja Mohan, *Crossing the Rubicon* (New York: Palgrave Macmillan, 2004).

sensitive nuclear transfers.[55] Nevertheless, for the time being, India's strategic conditions place it at risk of exporting sensitive nuclear material and technology to others.

To recap, an analysis of the outlier cases of Israel and India provides some support for the strategic theory of nuclear proliferation. While the outcome of the nuclear export behavior of these states is somewhat incongruent with the expectations of the strategic theory of nuclear proliferation, the within-case analysis demonstrates that the mechanisms identified by the strategic theory are clearly evident. In the end, even these "outlier" cases provide some support for the strategic theory of nuclear proliferation.

Pakistan to Iran, Libya, and North Korea, 1987–2002

The studies of a number of the historical cases presented to this point provide strong support for the strategic theory of nuclear proliferation, but how well can the theory travel to the present and explain post–Cold War cases? One may assume that in a world in which a single superpower is left to combat nuclear proliferation on its own, in which non-state actors are becoming increasingly important, and in which "proliferation rings" have the ability to flood the international market with sensitive materials and technology, the strategic theory of nuclear proliferation may no longer hold.[56] This assumption is incorrect. A brief analysis of a recent and important case provides strong support for the strategic theory of nuclear proliferation.

From 1987 to 2002, Pakistan, with the help of Pakistani scientist A. Q. Khan, exported sensitive nuclear materials and technologies to Iran, Libya, and North Korea.[57] Pakistan provided all three states with designs and key component parts

---

55. In 2005, the United States reversed its longstanding policy of opposition to India's nuclear program and signed a nuclear cooperation agreement with India. This type of nuclear assistance, the transfer of nonsensitive nuclear technologies to an existing nuclear power, is beyond the scope of this book's analysis. Still, one of key motivations of sensitive nuclear transfers also appears to be operative in this case. The United States was motivated to enhance India's nuclear capability in part by a desire to constrain China's growing power in Asia.

56. On proliferation rings, see, for example, Chaim Braun and Christopher Cyba, "Proliferation Rings," *International Security*, Vol. 29, No. 2 (Fall 2004), pp. 5–49.

57. The topic of Pakistan's nuclear exports has been the subject of a number of recent books that have uncovered much detail about the content, timing, and motivations of the sensitive nuclear exports. The best sources on this subject include Corera, *Shopping for Bombs;* Douglas Frantz and Kathleen Collins, *The Nuclear Jihadist* (New York: Twelve, 2007); William Langewiesche, *The Atomic Bazaar* (New York: Farrar Straus and Giroux, 2007); and Adrian Levy and Katherine Scott Clark, *Deception: Pakistan, the United States, and the Secret Trade in Nuclear Weapons* (New York: Walker and Company, 2007).

for uranium enrichment facilities. It also provided Libya (and perhaps the other states) with a nuclear weapon design. Initial reports claimed that the transfers were the result of a nuclear proliferation ring operated by a rogue Pakistani scientist without the support of the Pakistani government. It is now clear, however, as I will show, that the Pakistani state supported the policy of sensitive nuclear exports. This raises the critical question: Why did Pakistan export sensitive nuclear materials to Iran, Libya, and North Korea?

The strategic theory of nuclear proliferation provides a powerful explanation. Pakistan lacked the ability to project power over Iran and North Korea, meaning that Pakistan could provide sensitive nuclear assistance to these states without constraining its own military freedom of action. Pakistan was motivated to provide sensitive nuclear assistance in order to impose strategic costs on a state that threatened both Pakistan and the nuclear recipients: the United States. Finally, Pakistan did not depend on a superpower to provide for its own security, allowing it to continue with the sensitive nuclear transactions despite U.S. opposition. This case provides minimal support for rival explanations based on nuclear weapons possession, economics, institutions, substate smuggling, or technological quid pro quo.

Rogue Scientist or State Sponsorship?

Many of the relevant opinion-leaders have an incentive to write off Pakistan's nuclear transfers as being the result of a single rogue individual. The Pakistani government wishes to avoid responsibility for a voluntary act of nuclear proliferation, the U.S. government can comfortably continue counterterrorism cooperation with a key partner in the war on terror, and journalists can spin tales of a brilliant but mysterious scientist, engaged in intrigue and deception, whose actions singlehandedly transformed the landscape of international politics.

The truth is much less exciting and also—from the point of view of international relations theory—much less surprising. The sensitive nuclear exports of the A. Q. Khan network were state-sponsored by any reasonable definition of the term. Sensitive nuclear assistance is considered to be state-sponsored under two circumstances: if senior government officials explicitly authorize the assistance or if the assistance is conducted by a substate actor, but senior government officials are aware of the assistance and decline to stop it. In either situation it can be considered a de facto government policy.[58] Pakistan's sensitive nuclear exports qualify as state-sponsored *under both definitions*. It is now clear that the highest levels of Pakistan's government were aware of the transfers in the late 1980s

---

58. For a full discussion of state-sponsorship and sensitive nuclear assistance, see chapter 1.

but declined to stop them until 2002.[59] Moreover, as I will argue below, there is incontrovertible evidence that the government's support went much further and that senior Pakistani government officials explicitly authorized the sensitive nuclear transactions.

Pakistan is not the paragon of the social scientists' conception of a unitary-rational actor. Students of Pakistan often distinguish between the military and the Inter-Services Intelligence (ISI) on the one hand and the political elite on the other, as the often separate elements that make up the Pakistani state.[60] While this distinction is often important, it is less central in this case because, as I will show, both military and political leaders actively supported Pakistan's sensitive nuclear exports.

There is indisputable evidence that the Pakistani state actively supported and authorized the sensitive nuclear transactions. The government signed formal nuclear cooperation agreements with Iran, Libya, and North Korea. A. Q. Khan received official state support on the sensitive nuclear transfers. And senior government officials, including heads of state and military chiefs of staff, explicitly authorized the sensitive nuclear transfers.

The Pakistani government signed formal nuclear cooperation agreements with Iran, Libya, and North Korea, demonstrating that, at one point or another, the Pakistani government officially decided to help each of these countries develop their nuclear programs. In 1974, Prime Minister Zulfikar Ali Bhutto signed a nuclear cooperation arrangement with Libyan leader Colonel Muammar Al-Gaddafi.[61] Under the terms of the deal, Libya would help finance Pakistan's nuclear program, and in exchange, Pakistan would return sensitive nuclear materials and technology, including nuclear weapons, to Libya. The formal manifestation of this agreement was later cancelled, but the deal did facilitate subsequent transfers to Libya.

Pakistan's leadership also freely entered into a nuclear cooperation agreement with Iran. In 1987, the heads of the Pakistani and Iranian atomic energy commissions signed a secret nuclear cooperation agreement in Vienna.[62] This

---

59. On the government's role in the sensitive nuclear exports, see, Corera, *Shopping for Bombs*; Frantz and Collins, *Nuclear Jihadist*; Langewiesche, *Atomic Bazaar*; and Levy and Scott Clark, *Deception*.

60. On the influential role of the military in Pakistani politics see, for example, Ayesha Siddiqa, *Military Inc.* (New York: Pluto Press, 2007); Husain Haqqani, *Pakistan* (Washington, D.C.: Carnegie Endowment for International Peace, 2005); and Hassan Abbas, *Pakistan's Drift into Extremism* (New York: M.E. Sharpe, 2004).

61. For more information on the nuclear deal between Libya and Pakistan, see, for example, Corera, *Shopping for Bombs*, p. 12; and Frantz and Collins, *Nuclear Jihadist*, pp. 21–23.

62. On the Iranian-Pakistani nuclear cooperation agreement, see Farzad Bozoft, "Iran Signs Secret Atom Deal," *The Observer*, June 12, 1988.

formal arrangement between the two countries would help lay the groundwork for the sensitive nuclear exchanges between Pakistan and Iran that began later that year.

Pakistan also struck a formal nuclear cooperation deal with North Korea. In December 1993, Prime Minister Benazir Bhutto flew to Pyongyang, where she and North Korean dictator Kim Il Sung finalized the details of a trade in strategic technology. Under the terms of the deal, Pakistan would provide sensitive nuclear assistance to North Korea in exchange for North Korean missile technology. In a speech given at a banquet in Pyongyang, Bhutto announced that "Asian countries should cooperate with each other to develop their potentials" and that "Pakistan firmly holds the view that nuclear nonproliferation should not be made a pretext for preventing states from exercising fully their right to acquire and develop nuclear technology."[63]

The manner in which the transfers were conducted further demonstrates that A. Q. Khan was able to draw on steady government support. A. Q. Khan received official state support on the advertising, logistics, shipping, and security of the sensitive nuclear exports. Pakistan's Ministry of Commerce took out full-page ads in international newspapers advertising the sale of nuclear expertise and material.[64] Khan also appeared at international arms fairs as a representative of the Pakistani government, hawking sensitive nuclear technology. Brochures distributed by Khan Research Laboratories (KRL) advertised "nuclear-related products" and "complete ultracentrifuge machines."[65] Members of the Pakistani Foreign Ministry hosted Khan on his trips abroad to acquire materials for the nuclear supply program. Moreover, Khan delivered nuclear exports in military aircraft. For example, in 1998, Khan flew on a Pakistani Air Force Boeing 707 to North Korea, carrying five crates, containing P-1 and P-2 centrifuges, drawings, technical data, and uranium hexafluoride. The flight was approved by the military and chartered through official channels in Pakistan's Ministry of Defense. Beginning in 2000, Khan also delivered centrifuge components to Libya in official, government-chartered C-130 aircraft. On his travels abroad, the military even provided Khan with a security detail to protect him and his sensitive nuclear cargo.[66]

Moreover, senior government officials authorized and actively supported the sensitive nuclear transfers from 1987 to 2000. While none of these officials are willing to admit personal responsibility for the sensitive nuclear transfers, they

---

63. Benazir Bhutto, as cited in Corera, *Shopping for Bombs*, p. 86.
64. On the conduct of the sensitive nuclear transfers, see Corera, *Shopping for Bombs*, pp. 94–133; and Frantz and Collins, *Nuclear Jihadist*, pp. 204–213, 240.
65. Andrew Koch, "Khanfessions of a Proliferator," *Jane's Defense Weekly*, March 3, 2004.
66. See Corera, *Shopping for Bombs*, pp. 94–137.

are all eager to point fingers at others. When the accusations settle, the emergent picture suggests that everyone in the upper echelons of Pakistan's government played a role in Pakistan's nuclear exports. Every individual holding the powerful position of chief of army staff from 1987 to 2000 supported Pakistan's policy of nuclear transfer. According to A. Q. Khan, Pakistan's nuclear exports were fully authorized by Generals Zia, Beg, Karamat, Waheed, and Musharraf.[67] The involvement and motivations of some of these generals is now well understood. General Mirza Aslam Beg was vice chief of army staff from 1987 to 1988. Following President Zia's death in 1988, Beg became the chief of army staff (a position he would hold until 1991) and seventy-three-year-old Ghulam Ishaq Khan became president. As we will see below, Beg and President Khan authorized the nuclear transfers because they believed them to be essential to Pakistan's grand strategy. They believed that by promoting the spread of nuclear weapons, they could constrain U.S. military power and improve Pakistan's relative standing in regional affairs.

General Pervez Musharraf was also an advocate of the nuclear exports as a military officer in the 1990s and throughout his time as chief of army staff from 1998 to 1999. According to many accounts, Musharraf was willing to trade nuclear technology to North Korea in exchange for missile technology, because he believed that long-range missiles were necessary for Pakistan to deter an attack from neighboring India. His support for the transfers likely continued when he assumed power in 1999. Hendrina Khan, A. Q. Khan's wife, claims that President Musharraf issued direct orders to A. Q. Khan to provide sensitive nuclear assistance to North Korea.[68]

Musharraf claims that he was completely unaware of Pakistan's sensitive nuclear exports until he put a stop to them, but his denials do not stand up to scrutiny. In an attempt to prove his ignorance of the transfers, Musharraf claims that in late 2000, he summoned A. Q. Khan to his office to question him about rumors that KRL was exporting sensitive nuclear materials and technology.[69] According to Musharraf, Khan denied the accusations. But rather than citing any personal attempts to verify Khan's claims independently or shut down Khan's nuclear dealings, Musharraf writes in his memoirs simply that he "remained apprehensive."[70] Moreover, Musharraf's failure to stop the sensitive nuclear exports appears to have been a matter of volition rather than capability. After all, when he decided to shut down Khan's operations on January 31, 2004, he was

---

67. "Khan: Musharraf in on North Korea Nuke Deal," *Associated Press*, July 4, 2008; and Corera, *Shopping for Bombs*, p. 96.

68. "Musharraf Lied about A.Q.'s Role in N-Program Says Hendrina," *Daily Times* (Pakistan), August 10, 2008.

69. For Musharraf's brief description of Pakistan's sensitive nuclear exports during his presidency, see Pervez Musharraf, *In the Line of Fire* (New York: Free Press, 2006).

70. Ibid.

able to do so quite easily. Following Libya's disclosures of its nuclear dealings with Khan in December 2003 and facing considerable international pressure, President Musharraf quickly placed Khan under house arrest. By not stopping Khan sooner, Musharraf, at a minimum, implicitly endorsed Khan's activities.

While Musharraf is unwilling to accept responsibility for the nuclear exports, he is perfectly willing to place the blame on others. Musharraf claims that, throughout the 1990s, Pakistan's sensitive nuclear exports were being run out of the president's office.[71] As demonstrated above, it is clear that President Khan authorized the sensitive nuclear transfers while he held power from 1988 to 1993. If Musharraf's accusations are correct, however, the Pakistani presidents that are responsible for the sensitive nuclear transfers also include Wasim Sajjad (1993, 1997–1998), Farooq Leghari (1993–1997), and Muhammad Rafiq Tarar (1998–2001).

Benazir Bhutto also actively supported the sensitive nuclear exports while serving as prime minister of Pakistan from 1993 to 1996. Bhutto flew to Pyongyang to finalize a trade with North Korea involving nuclear technology. In fact, Shyam Bhatia, a close friend of Prime Minister Bhutto and a respected British journalist, claims that Bhutto told him that she carried the uranium centrifuge designs to North Korea on a CD in her own coat pocket and handed them to Kim Il Sung personally.[72] Bhutto justified her support for the sensitive nuclear transfers in a remark to an aide thus: "I thought the military would be very happy with me and would stop trying to destabilize my government."[73]

U.S. officials working on nuclear proliferation issues also believe that Pakistan's sensitive nuclear transfers were state-sponsored. According to Gary Samore, senior director for nonproliferation in the White House from 1996 to 2001, "The Pakistani government were probably aware of and supported the transfers."[74]

In sum, there is overwhelming support for the idea that senior political and military leaders in Pakistan supported a policy of sensitive nuclear transfer. This is certainly more plausible than the alternative explanation that a single individual exported Pakistan's most sensitive nuclear technology for over a decade without anyone in a position of power knowing about it or approving it.

Pakistan's Power Projection

Pakistan's nuclear exports were generally channeled to states over which Pakistan lacked the ability to project conventional military power. This meant that Pakistan could provide sensitive nuclear assistance to these states without constraining its own military freedom of action.

---

71. Corera, *Shopping for Bombs*, p. 214.
72. Shyam Bhatia, *Goodbye Shahzadi* (New Delhi: Eastern Book Corporation, 2008).
73. Benazir Bhutto, as cited in Frantz and Collins, *Nuclear Jihadist*, p. 207.
74. Corera, *Shopping for Bombs*, p. 78.

Pakistan clearly lacked the capabilities that would have allowed it to project military power over Libya and North Korea.[75] These countries are geographically distant from Pakistan, making a ground invasion of these countries impossible. Pakistan also lacked the amphibious invasion or airlift capabilities that would have allowed it to project power against these distant states.

Because Pakistan lacked the ability to project power against these states, Pakistani officials were not worried that nuclear proliferation in these cases would undermine Pakistan's own military freedom of action. For example, when asked how the acquisition of nuclear weapons by Iran, Libya, or North Korea would affect Pakistan's own security, Jehangir Karamat, Pakistan's Ambassador to the United States from 2004 to 2006 and former chief of army staff from 1996 to 1998, admitted, "North Korean nuclear capability does not threaten us directly."[76]

Iran is a different story. Iran and Pakistan share a long border, and Pakistan could conceivably project power into Iran. By helping Iran to acquire nuclear weapons, Pakistan did potentially risk constraining its own military freedom of action. This was an outcome recognized by Pakistani officials. For example, Ambassador Karamat also acknowledged the potential security problems that a nuclear-armed Iran could bring to Pakistan.[77]

The strategic theory of nuclear proliferation, like most social science theories, is a probabilistic theory, not a deterministic one. The argument of Hypothesis 1 is not that states will never provide sensitive nuclear assistance to states over which they can project military power but that they will be more likely to provide sensitive nuclear assistance to states over which they lack power-projection capabilities. In the case of Pakistan's nuclear transfers, we find support for this probabilistic statement. In two of the three cases in which Pakistan exported sensitive nuclear materials and technology, the recipient was a state over which Pakistan lacked the ability to project military power.

Pakistan's Common Enemies

Pakistan provided sensitive nuclear assistance to Iran, Libya, and North Korea in order to impose strategic costs on a common rival, the United States.

Iran, Libya, and North Korea were all locked in a state of enmity with the United States and were all threatened by American military power. The United

---

75. On the capabilities of the Pakistani military, see, for example, Ayesha Siddiqa-Aqha, *Pakistan's Arms Procurement and Military Build-Up, 1979–1999* (New York: Palgrave Macmillan, 2001).

76. Karamat also made it clear that nuclear proliferation in North Korea could cause problems in East Asia and that Pakistan firmly opposes the international spread of nuclear weapons. Interview with the author, April 2006.

77. Ibid.

States and Iran have been enemies since the 1979 Iranian Revolution in which Iranian students seized the U.S. Embassy and held fifty-two American hostages captive for 444 days.[78] A U.S. military intervention into Iran to rescue the hostages failed and further exacerbated U.S.-Iranian relations. The United States and Libya have also been at odds since General Muammar Al-Gaddafi seized power in 1969.[79] While in charge of Libya, Gaddafi closed U.S. military bases in the country, adopted an anti-Western foreign policy orientation, and sponsored incidents of terrorism against the West. The United States responded with missile strikes on Gaddafi's residence that narrowly missed the intended target and killed his daughter. The United States and North Korea have also been enemies since the end of World War II.[80] The Korean War brought the United States and Communist North Korea face-to-face in direct military combat. Following the collapse of the Soviet Union, the Communist government's dogged pursuit of nuclear weapons continually antagonized Washington. In 1994, the U.S. government seriously considered a preventive military strike against North Korea's nuclear facilities. Indeed, in the 2002 State of the Union Address, President Bush recognized two of these three states as Washington's worst enemies, labeling Iran and North Korea as part of an "axis of evil."[81]

In the 1990s, Pakistan was also threatened by American military power. Although the United States had sometimes engaged Pakistan in a partnership of strategic convenience, the two had never been close allies. The United States had aided Pakistan in its efforts to eject Soviet forces from Afghanistan in the 1980s, but once the Soviet threat receded, the United States quickly ended its support to Pakistan.[82] The Pakistani elite felt that they had been abandoned by the United States. Moreover, they were also threatened by the power and influence that Washington now possessed as the world's only remaining superpower. For example, General Beg believed that, following the collapse of the Soviet Union, the United States had become the dominant force in South Asia and the Middle East and that U.S. power could potentially threaten Pakistan's interests.[83]

---

78. For a recent treatment of U.S.-Iranian relations, see Trita Parsi, *Treacherous Alliance* (New Haven: Yale University Press, 2007).

79. On U.S.-Libyan relations, see, for example, Ronald Bruce St. John, *Libya and the United States* (Philadelphia: University of Pennsylvania Press, 2002).

80. On U.S.-North Korean relations, see, for example, Victor D. Cha and David C. Kang, *Nuclear North Korea* (New York: Columbia University Press, 2003); and Charles L. Pritchard, *Failed Diplomacy* (Washington, D.C.: Brookings Institution Press, 2007).

81. "The President's State of the Union Address," January 29, 2002.

82. On U.S.-Pakistani relations, see, for example, Levy and Scott-Clark, *Deception*.

83. On Pakistani concerns about a unipolar world dominated by the United States and the idea that the spread of nuclear weapons to additional states could contain U.S. power, see, for example, Corera, *Shopping for Bombs*, pp. 74–79, 122–124; MSNBC Interview with Mirza Aslam Beg, February 9, 2004; Haqqani, *Pakistan*, p. 280; and Frantz and Collins, *Nuclear Jihadist*, p. 159.

Pakistan's sensitive nuclear exports were motivated in part to constrain U.S. military power.[84] By helping a band of pariah states acquire nuclear weapons, senior Pakistani officials hoped to impose strategic costs on the United States. General Beg thought that the increased global diffusion of nuclear weapons could lead to a multipolar world that would better suit Pakistan's interest than a bi- or unipolar world dominated by the United States. Beg believed that Pakistan and a group of states hostile to Washington, armed with nuclear weapons and supported by China, could form an alliance of "strategic defiance" against the United States. With the United States forced out of the region, Pakistan and its allies could assume a greater leadership role in the Middle East and South Asia. A. Q. Khan was also motivated by a desire to undermine the strategic interests of the United States.[85] Khan frequently told friends that he enjoyed providing sensitive nuclear assistance to countries that were opposed to the United States. In speeches to international audiences, Khan also preached the message that nuclear proliferation was the best antidote to too much American power. For example, in a speech in Syria in the 1990s, Khan argued that Syria should acquire nuclear weapons in order to protect itself from the United States. By the time the sensitive nuclear transfers came to an end, Khan appears to have believed that he had achieved his goal. He proudly proclaimed, referring to the United States, "I disturbed all their strategic plans, the balance of power and blackmailing potential in this part of the world."[86]

In sum, according to Gordon Corera, Pakistan provided sensitive nuclear assistance because key Pakistani leaders believed that it was in "Pakistan's national interest for more countries to have bombs, thereby... reducing the power of the United States."[87]

Pakistan's Superpower Dependence

The United States, a superpower, was threatened by nuclear proliferation anywhere on the globe and intervened in an attempt to stop Pakistan's nuclear transfers. Pakistan did not depend on the United States to provide for its own security, allowing it to continue the sensitive nuclear transactions despite U.S. opposition.

---

84. On the belief in Pakistan that the promotion of nuclear proliferation could contain U.S. power and improve Pakistan's security, see ibid.
85. On Khan's anti-Americanism, see, for example, Corera, *Shopping for Bombs*, pp. 122–124; and Frantz and Collins, *Nuclear Jihadist*, pp. 211, 223.
86. Zeba Khan, Interview, "Abdul Kadeer Khan: The Man Behind the Myth."
87. Corera, *Shopping for Bombs*, p. 79.

The United States was concerned about the strategic effects of Pakistan's nuclear activities and intervened in an attempt to stop them.[88] In the fall of 1990, in response to President George H. W. Bush's inability to verify that Pakistan did not possess nuclear weapons, the United States imposed sanctions and cut off all arms transfers to Pakistan, including the sale of twenty-eight American F-16 fighter aircraft. In 1998, following Pakistan's nuclear tests, the Clinton administration expanded sanctions against Islamabad. Under the expanded sanctions, the United States blocked any loans to Pakistan from the U.S. government, U.S. banks, the International Monetary Fund (IMF), and the World Bank. In 2003 the United States placed sanctions directly on KRL.

The United States also employed its powerful intelligence agencies to thwart Pakistan's nuclear exports. The CIA attempted to undermine the value of Pakistan's nuclear exports by using middlemen to pass on flawed nuclear technology to Pakistan's nuclear recipients. U.S. officials even considered assassinating A. Q. Khan to stop his nuclear exploits.[89]

The United States also applied strong diplomatic pressure to stop Pakistan's sensitive nuclear exports. In January 1990, U.S. Assistant Secretary of Defense Henry Rowen warned General Beg that if Pakistan exported sensitive nuclear materials and technology, "Pakistan would be in serious trouble with the United States."[90] The Clinton administration would continue to issue letters of complaint and express its concerns about the nuclear exports with Pakistani officials throughout the 1990s.

Pakistan resisted these pressures and continued with the sensitive nuclear transactions until 2003. In fact, Pakistan's resistance followed a pattern. When confronted by U.S. officials, the Pakistanis would claim that they were unaware of any nuclear transfers, that U.S. accusations were not specific enough for them to take any action, and that they would look into the problem. But Pakistani officials did not keep their promises to examine the issue, and they did not halt the sensitive nuclear exports. For example, when President G. I. Khan was warned of A. Q. Khan's nuclear exports in 1989, G. I. Khan summoned A. Q. Khan to his office. Instead of ordering him to stop the transfers, he advised him to "be careful."[91] When Bill Clinton raised the nuclear export issue with Prime Minister Nawaz Sharif at the White House in 1998, Sharif responded that "he certainly was not aware of such activity, but would look into it on his return to Islamabad."[92]

---

88. On U.S. efforts to stop Pakistan's nuclear exports, see, for example, Correra, *Shopping for Bombs*, pp. 80–216; and Frantz and Collins, *Nuclear Jihadist*, pp. 89–242.
89. See Frantz and Collins, *Nuclear Jihadist*, pp. 89, 250.
90. Henry Rowen, as quoted in Frantz and Collins, *The Nuclear Jihadist*, p. 177.
91. Corera, *Shopping for Bombs*, p. 96.
92. Nawaz Sharif, as quoted in Corera, *Shopping for Bombs*, p. 139.

On his return to Islamabad, he took no specific action. In January 1999, U.S. Deputy Secretary of State Strobe Talbott flew to Islamabad to convince Sharif to halt the nuclear transfers. Sharif admitted that Pakistan and North Korea had cooperated on conventional military technology but denied any nuclear-related exchanges. Similarly, as was detailed above, when Musharraf assumed the office of president, he called Khan into his office for questioning—he claims—but did not press any further after receiving Khan's denial.

Pakistan eventually put an end to its sensitive nuclear transfers. In December 2003, after Libya disclosed that it had received sensitive nuclear technology from Pakistan and after U.S.-Pakistani relations had found a better footing due to Pakistan's cooperation in the U.S.-led War on Terror, the United States finally gained the leverage it needed to convince Pakistan to shut down the nuclear proliferation ring. On January 31, 2004, Musharraf placed Khan under house arrest. By then, however, the damage had been done. For over a decade, Pakistan had managed to export sensitive nuclear materials and technology to Iran, Libya, and North Korea despite U.S. opposition.

Pakistan was able to resist U.S. pressure on the nuclear export issue for so long because it did not depend on the United States to provide for its security. Pakistan is believed to have assembled its first nuclear weapon in about 1990, granting it superpower independence. Moreover, Pakistan did not in this period have a security guarantee from Washington and did not enjoy a place under the American nuclear umbrella. Because Pakistan did not rely on the United States to provide for its security, it could more easily ignore Washington's protests and continue with the sensitive nuclear exports without the concern that they might antagonize a powerful patron.

One may also question whether the United States really put serious pressure on Pakistan to stop the nuclear transfers or whether it was perhaps all too willing to turn a blind eye to A. Q. Khan's nuclear dealings. This line of objection is understandable, but mistaken. While it is true that Washington did not always make proliferation the foremost issue in its bilateral relations with Pakistan, the argument that the United States did not seriously pressure Pakistan to stop its nuclear transfers is untenable and may arise from an unrealistic expectation about the feasible policy options that Washington had at its disposal. It may be possible to imagine more extreme forms of coercion that Washington might have hypothetically brought to bear against Pakistan, but there is no question that, from the first knowledge of its existence, the United States employed an entire range of pragmatic diplomatic and intelligence tools in an attempt to monitor and then shut down the A. Q. Khan network.

It could also be argued that while Pakistan was not superpower-dependent according to a strict definition of the term, it was nevertheless dependent on the United States. The United States has historically been one of Pakistan's largest arms suppliers, and Pakistan and the United States have often engaged in a partnership of convenience. While, this may be true, the on-again/off-again strategic partnership between the United States and Pakistan was decidedly off during the 1990s, when the nuclear transfers occurred. The transfers occurred after the United States and Pakistan had ended their cooperation to fight the Soviet Union in Afghanistan during the 1980s but before Washington and Islamabad renewed their strategic partnership after the terrorist attacks on September 11, 2001, to fight the War on Terror. During the 1990s, Pakistan did not view Washington as a reliable ally, the U.S.-Pakistan relationship was at a decided low, and Pakistan had nothing to lose by ruffling Washington's feathers. At various times, Pakistan may have been concerned about the U.S. reaction to its nuclear exports, but it did not depend on the United States to provide for its security, allowing it to continue with the sensitive nuclear transactions in the face of Washington's protests.

In sum, Pakistan's sensitive nuclear assistance to Iran, Libya, and North Korea provides strong support for the strategic theory of nuclear proliferation. In support of Hypothesis 1, Pakistan was more likely to provide sensitive nuclear assistance to states over which it lacked the ability to project military power. Consistent with Hypothesis 2, Pakistan provided sensitive nuclear assistance to states that were, like Pakistan, threatened by the United States. And, in support of Hypothesis 3, the United States intervened in an attempt to prevent Pakistan's sensitive nuclear exports, but Pakistan was able to continue the transactions because it was not dependent on a superpower.

Alternative Explanations

The strategic theory of nuclear proliferation provides a more compelling explanation of Pakistan's sensitive nuclear exports to Iran, Libya, and North Korea than alternative accounts that focus on economics, institutions, nuclear weapons possession, smuggling, or strategic quid pro quo.

ECONOMICS

An economic explanation falls short. While some individuals involved in the sensitive nuclear transfers may have been motivated by economic gain, the sums of money the Pakistani government received for its sensitive nuclear exports were simply too small to affect a state's calculations.

It is estimated that Libya paid between $80 to $100 million in total for the nuclear assistance it received from Pakistan.[93] However, much of these funds were spent on the procurement and transfer of the nuclear cargo, and the rest was divided up among A. Q. Khan and other key actors in the network. None of this currency made its way back to Islamabad, raising serious doubts as to whether the Pakistani government could have reasonably expected to receive a substantial payment in return for its sensitive nuclear transfers. Moreover, even if the Pakistani government expected to receive the entire sum, it would not have been a sufficient reward to justify such a potentially risky foreign policy decision. It is simply implausible that a country would consciously trade away its own security for a mere $100 million. This suggests that Pakistani officials saw the exports as consistent with their security goals.

In the cases of Iran and North Korea, an economic explanation is even less plausible. In Iran, the cash exchanged in the nuclear transactions was trivial. Iran paid Pakistan a token amount, estimated at between $3 to $10 million for its sensitive nuclear assistance.[94] This would be enough money to pay off a few individuals handsomely but would not so much as rattle the change in the state's coffers. Finally, economic concerns were completely absent in the case of the exports to North Korea. Pakistan did not receive any hard currency in exchange for its sensitive nuclear exports to North Korea. Instead, North Korea provided Pakistan with missile technology.

INSTITUTIONS

Institutional arguments find some support in this case. Pakistan has never joined the NPT or the NSG. This fact is consistent with the expectation that countries that are members of the institutions of the nuclear nonproliferation regime will be less likely to provide sensitive nuclear assistance. Arguments about the constraining effects of international norms on nuclear proliferation, however, are not supported. Proponents of the norm-based argument often contend that the international norm against nuclear proliferation was established alongside the NPT in 1968 and that the norm has strengthened over time as more and more countries sign the treaty. If this argument is correct, the norm against nuclear proliferation should have been very strong by the 1990s, as membership in the NPT reached near-universality. Yet, it was during this exact time period, from 1987 to 2002, that Pakistan continually exported sensitive nuclear materials and technology to Iran, Libya, and North Korea. If global norms against nuclear

---

93. For these estimates, see Corera, *Shopping for Bombs*, p. 120; and Frantz and Collins, *Nuclear Jihadist*, p. 319.

94. Corera, *Shopping for Bombs*, pp. 60, 66; and Frantz and Collins, *Nuclear Jihadist*, pp. 158–161.

proliferation were stronger than ever at the turn of the century, they were not strong enough to prevent the most egregious case of sensitive nuclear assistance in history.

NUCLEAR WEAPONS POSSESSION

The argument that nuclear weapon states are less likely to provide sensitive nuclear assistance does not find support in this case. Pakistan exported sensitive nuclear materials to Iran, Libya, and North Korea at a time when it possessed nuclear weapons itself. Pakistan did not seem intent on preventing other states from joining the nuclear club. Instead, it acted as a hyperactive club promoter, handing out free V.I.P. memberships. This evidence further undermines arguments about nuclear weapon states being particularly opposed to nuclear proliferation.

SUBSTATE SMUGGLING

A smuggling argument finds no support in this case. Although not one of the alternative explanations for the supply-side of nuclear proliferation presented in chapter 1, some have suggested that Pakistan's nuclear exports were the result of substate smuggling. According to this narrative, A. Q. Khan and the cronies in his network sold Pakistan's sensitive nuclear materials and technology abroad for their own personal gain, without the knowledge of the Pakistani state. The popular press story of a rogue scientist who was motivated by greed and megalomania and operated without state sanction has already been thoroughly discredited. As has been demonstrated above, the Pakistani state clearly authorized and supported all of the sensitive nuclear transactions under discussion.

STRATEGIC QUID PRO QUO

A strategic quid pro quo argument cannot explain Pakistan's sensitive nuclear exports either. While also not one of the alternative explanations for sensitive nuclear assistance presented in chapter 1, some descriptions of the A. Q. Khan case have seized on the importance of the missile technology that North Korea gave to Pakistan in return for sensitive nuclear assistance. Extrapolating from this case, one could hypothesize that states provide sensitive nuclear assistance in exchange for other valuable military technologies. This hypothesis would certainly find support in the case of Pakistan's assistance to North Korea, but when one considers Pakistan's other nuclear exports to Iran and Libya, the explanation falls short. Neither Iran nor Libya provided Pakistan with strategic technology in return for nuclear cooperation. Furthermore, the strategic quid pro quo between Pakistan and North Korea appears to be a unique event. There are no other identifiable cases in which a nuclear supplier provided sensitive nuclear assistance in exchange for another strategic military technology.

Conclusion

In this chapter we have seen that the strategic theory of nuclear proliferation receives strong support in a broad set of cases. The "enemy of my enemy is my customer" logic is not unique to the France-Israel case. Rather, an analysis of China's sensitive nuclear assistance to Pakistan from 1981 to 1986 demonstrates that the desire to constrain a common enemy has encouraged sensitive nuclear assistance in other contexts.

The strategic theory of nuclear proliferation cannot perfectly explain the details of every case of sensitive nuclear assistance, but it still finds support even in a series of outlier cases. These outlier cases include states that should not have provided nuclear assistance but did (such as the Soviet Union) and countries that should have provided sensitive nuclear assistance but did not (such as Israel and India). The strategic theory of nuclear proliferation cannot account for all aspects of the Sino-Soviet case, but it provides a firmer basis for understanding than competing explanations. An analysis of the Israel and India cases demonstrates that the strategic factors motivating nuclear assistance are present in these cases and may still convince these countries to export sensitive nuclear materials and technologies. Although, if India's military power continues to grow, New Delhi may become less likely to export sensitive nuclear materials over time.

Finally, this chapter examined Pakistan's nuclear assistance to Iran, Libya, and North Korea. These cases demonstrate that, far from being a theory that is applicable only to the Cold War, the strategic theory of nuclear proliferation provides us with a helpful framework for understanding sensitive nuclear assistance in the post–Cold War world.

The case studies presented in this chapter examine discrete occurrences of sensitive nuclear assistance, but the strategic theory of nuclear proliferation can also shed light on changes in sensitive nuclear export behavior over time. As the conditions identified by the strategic theory of nuclear proliferation change, states' incentives to provide sensitive nuclear assistance should vary accordingly. As a state's power projection capabilities rise or fall, as a shared enemy appears or disappears, and as a state becomes more or less dependent on a superpower patron, the probability of sensitive nuclear assistance also changes.

The ability of the strategic theory of nuclear proliferation to explain changes in sensitive nuclear export behavior over time can be seen in a brief examination of nuclear export policy in Russia and China. In recent years, Russia has betrayed an increased willingness to export sensitive nuclear materials and technology.

During the Cold War, it could be argued that no state enforced a tougher nuclear nonproliferation policy than the Soviet Union.[95] The Soviet Union was a key player in the establishment of the institutions of the nuclear nonproliferation regime. It was often willing to consider tough measures, including the use of military force, to prevent other countries from acquiring nuclear weapons. And, with the exception of a brief period of half-hearted sensitive nuclear transfers to China, the Soviet Union never exported sensitive nuclear materials and technologies to a nonnuclear weapon state.

Following the collapse of the Soviet Union, however, Russia appears to be much less concerned about combating nuclear proliferation. Indeed, by 1995, just five years after the collapse of the Soviet Union, Russia had become a major exporter of nuclear technology. Russia entered into a contract with Iran to finish the construction of two nuclear power reactors at Bushehr.[96] More strikingly, Russia even offered to provide Iran with a sensitive uranium-enrichment facility. Russia reluctantly cancelled the uranium enrichment deal under intense U.S. pressure, but that Russia was willing to provide sensitive nuclear assistance to Iran signifies a major shift in thinking on nuclear proliferation issues in Moscow. Some analysts even speculate that Moscow's continued civilian nuclear and military cooperation with Iran may be driven by an "enemy of my enemy is my customer" logic and a desire to constrain U.S. power.[97]

China is on the opposite trajectory. Traditionally, China followed an extremely lax nuclear nonproliferation policy. As was stated above, China was unwilling to join the NPT and the NSG for decades. China also repeatedly provided sensitive nuclear assistance to nonnuclear weapon states. From 1981 to 1986, China exported sensitive nuclear materials and technology to Pakistan. At various times between 1984 and 1995, China provided sensitive nuclear aid to Iran. And from 1986 to 1991, China helped Algeria build sensitive nuclear fuel-cycle facilities.

Yet, in recent years, China has performed an about-face on nuclear proliferation issues. China joined the NPT in 1992 and the NSG in 2004. It has cut off its sensitive nuclear dealings with Iran and Algeria and has not entered into any new sensitive nuclear export agreements. Moreover, Beijing is assuming an active role in international diplomacy on high-profile nuclear proliferation

---

95. On the Soviet Union's nuclear nonproliferation policy see, for example, William C. Potter, "U.S.-Soviet Cooperative Measures," in Jones et al., eds., *Nuclear Suppliers and Nonproliferation*.
96. On Russian-Iranian nuclear cooperation, see, for instance, Orlov and Vinnikov, "Great Guessing Game."
97. See, for example, Robert A. Pape, "Soft Balancing against the United States," *International Security*, Vol. 30, No. 5 (Fall 2005), pp. 7–45.

problems. Most notably, China hosted the Six-Party Talks aimed at convincing North Korea to abandon its nuclear weapons program.[98]

What explains the divergent trajectories of Russia and China on nuclear proliferation issues? Why is China becoming less willing to provide sensitive nuclear assistance at the same time that Russia is becoming more willing to do so?

Undoubtedly, there are many factors that may have a role to play in these outcomes. Nevertheless, the strategic theory of nuclear proliferation presented in this book provides a partial answer to this question. During the Cold War, the Soviet Union was a superpower with global force-projection capabilities such that nuclear proliferation anywhere constrained its military power. For this reason, the Soviet Union took a tough stand against nuclear proliferation and was loath to provide sensitive nuclear assistance. But Moscow's conventional military power collapsed along with the Soviet Union and, therefore, nuclear proliferation no longer threatened to constrain Moscow's force-projection capability in the same way.[99] Partly for this reason, Russia appears to be more willing to provide sensitive nuclear assistance.

China is on a very different path. China's power-projection capability has traditionally been limited to a handful of states on its immediate borders. For most of the nuclear era, nuclear proliferation to states beyond this sphere of influence did not directly threaten to constrain China's military power. For this reason, China could provide sensitive nuclear assistance beyond this sphere of influence without worrying about the strategic consequences. In recent years, however, as China modernizes its military forces and begins to think about projecting power to distant regions, it is becoming more concerned that nuclear proliferation could constrain its own power. This may be one consideration making China less willing to provide sensitive nuclear assistance. Though the rise of Chinese power may pose a threat to U.S. security interests in certain domains, it may actually aid U.S. nonproliferation efforts. A Chinese superpower could become a partner with which the United States could work to combat the international spread of nuclear weapons.[100] Conventional military weakness in Russia, on the other hand, could be a factor that encourages Moscow to help other countries fulfill their nuclear ambitions.

---

98. On the Six-Party Talks, see Carin Zissis, "The Six-Party Talks on North Korea's Nuclear Program," *Council on Foreign Relations*, June 2008.

99. On the collapse of Russian military power following the collapse of the Soviet Union, see, for example, Dunnigan, *How to Make War*, p. 13.

100. For a more detailed version of the argument that greater numbers of superpowers may lead to a reduction in transnational security threats such as nuclear proliferation, see Steven Weber, Naazneen Barma, Matthew Kroenig, Ely Ratner, "How Globalization Went Bad," *Foreign Policy*, Vol. 86, No. 1 (January/February 2007), pp. 48–54.

5

IMPORTING THE BOMB
Nuclear Assistance and Nuclear Proliferation

In previous chapters, I explained the causes of sensitive nuclear assistance. But why should we care about the causes of sensitive nuclear assistance? Does sensitive nuclear assistance lead to the proliferation of nuclear weapons? The idea that states that get help with their nuclear programs will be more likely to acquire nuclear weapons has intuitive appeal, but international nuclear transfers may have no meaningful effect on nuclear proliferation. States with an intense demand for nuclear weapons or sufficient domestic industrial capabilities may be able to acquire nuclear weapons whether they receive sensitive nuclear assistance or not. Similarly, states that lack the desire to build nuclear arms or the requisite industrial capacity may not produce a nuclear arsenal even with substantial foreign aid. Indeed, existing scholarly approaches to nuclear proliferation have examined why states want nuclear weapons as well as the relationship between domestic capacity and nuclear acquisition, but none has explored whether or not international nuclear assistance affects nuclear proliferation.[1] This raises an important question about the sources of nuclear proliferation: Does sensitive nuclear assistance contribute to the international spread of nuclear weapons?

To answer this question, I begin with a simple logic of the technical and strategic advantages that potential nuclear proliferators can gain by importing sensitive nuclear materials and technologies from more advanced nuclear states.

---

1. For previous approaches to the causes of nuclear proliferation, see, for example, Sagan, "Why Do States Build Nuclear Weapons?"; Singh and Way, "Correlates of Nuclear Proliferation"; and Jo and Gartzke, "Determinants of Nuclear Weapons Proliferation."

I argue that states that receive sensitive nuclear assistance can better overcome the common obstacles that states encounter as they attempt to develop a nuclear weapons arsenal. They can leapfrog technical design stages, acquire tacit knowledge from more advanced scientific communities, economize on the costs of nuclear development, and avoid international pressure to abandon a nuclear program.

Drawing on the new data on the international transfer of sensitive nuclear materials and technology presented earlier in the book, this chapter demonstrates that sensitive nuclear assistance is an important determinant of nuclear proliferation. States that receive sensitive nuclear assistance from abroad are more likely to acquire nuclear weapons. Sensitive nuclear assistance remains an important predictor of nuclear proliferation even after controlling for state demand for nuclear weapons and domestic industrial capacity. I find no support for the idea that nonsensitive, civilian nuclear cooperation contributes to the spread of nuclear weapons. I find modest support for the idea that a threatening security environment is an important predictor of nuclear proliferation.

Although its key contribution is in providing empirical evidence in support of a new explanation for nuclear proliferation, this chapter also demonstrates the importance of focusing on the supply side of nuclear proliferation. Existing scholarly approaches to nuclear proliferation have thoroughly examined why states want nuclear weapons. By studying the effect of international flows of nuclear material and technology on nuclear proliferation, this chapter shifts attention from the demand for nuclear weapons to their supply.

Explaining Nuclear Proliferation

There is a vast scholarly literature on the causes of nuclear proliferation. Dong-Joon Jo and Erik Gartzke have recently categorized this research into two camps: arguments that focus on a state's *willingness* to acquire nuclear weapons (demand-side approaches) and those that privilege a state's *opportunity* to acquire nuclear weapons (supply-side approaches).[2] The vast majority of scholarly research on nuclear proliferation has focused on demand. This school has sought to identify the factors that drive states to pursue and abandon nuclear weapons programs. Scott Sagan argues that there are three primary reasons why states seek nuclear weapons.[3] Sagan maintains that states in competitive security environments desire nuclear weapons as a means to deter external aggression, that domestic

---

2. Jo and Gartzke, "Determinants of Nuclear Weapons Proliferation."
3. Sagan, "Why Do States Build Nuclear Weapons?"

political lobbies (primarily the domestic nuclear complex) can encourage states to pursue a national nuclear weapons program for parochial reasons, and that international norms of prestige or opprobrium associated with nuclear weapons can influence states' nuclear decisions. Sagan concludes that none of these causes is dominant but that they each operate to varying degrees in different cases.

Other scholars have suggested additional factors that may influence a state's demand for nuclear weapons. Etel Solingen maintains that domestic political coalitions and their associated economic development strategies determine a state's demand for nuclear weapons.[4] "Liberalizing coalitions" are internationalist, pursue export-oriented industrialization strategies, and will be reluctant to jeopardize international trade and investment on controversial foreign policies, such as the pursuit of nuclear weapons. On the other hand, states controlled by "inward looking, nationalist, and radical-confessional coalitions" oppose liberalization and choose an autarchic path to economic development. According to Solingen, they are more likely to pursue nuclear weapons because they face fewer international economic costs to doing so and are more beholden to nationalist appeals.

Individual psychological drivers have also been invoked to explain a state's willingness to acquire nuclear weapons. Jacques Hymans argues that leaders' conceptions of their countries' national identities is the key to explaining state demand for nuclear weapons.[5] Other research has drawn on these and other factors to explain why states pursue and abandon nuclear weapons programs.[6]

In contrast, the supply-side approach to nuclear proliferation recognizes that an analysis of a state's demand for nuclear weapons can only provide a partial explanation for nuclear proliferation.[7] Whether or not a state wants nuclear weapons is irrelevant if it is unable to acquire them. States may badly desire nuclear weapons but lack the technology, resources, and expertise required to build them. Moreover, opportunity can shape willingness. States that could conceivably produce a nuclear weapons arsenal will face a great temptation to go nuclear. According to this view, "Once a country acquires the latent capacity to

---

4. Etel Solingen, "The Political Economy of Nuclear Restraint," *International Security*, Vol. 19, No. 2 (Fall 1994), pp. 126–169; and Solingen, *Nuclear Logics*.

5. Jacques E. C. Hymans, *The Psychology of Nuclear Proliferation* (Cambridge: Cambridge University Press, 2006).

6. See, for example, Paul, *Power Versus Prudence*; and George Quester, *The Politics of Nuclear Proliferation* (Baltimore: Johns Hopkins University Press, 1973).

7. For scholarship that focuses on the relationship between domestic capacity and nuclear proliferation, see, for example, Singh and Way, "Correlates of Nuclear Proliferation"; Jo and Gartzke, "Determinants of Nuclear Proliferation"; Stephen M. Meyer, *The Dynamics of Nuclear Proliferation* (Chicago: University of Chicago Press, 1984).

develop nuclear weapons, it is only a matter of time until it is expected to do so."[8] The supply-side approach to proliferation claims that states with an advanced industrial capacity can more easily create and maintain a nuclear weapons program and are thus more likely to acquire nuclear weapons than less-developed states. This line of reasoning has roots in earlier scholarship and has been revived by recent quantitative analyses of nuclear proliferation.[9] The quantitative studies have found that measures of economic development and industrial capacity are associated with a greater risk of becoming a nuclear power. These authors do not consider, however, how international nuclear assistance may advance a country's opportunity to produce nuclear weapons, nor do they explicitly examine the relationship between international nuclear assistance and nuclear proliferation.

The literature on "proliferation rings" has argued that nuclear capabilities in "second-tier" supplier states like Pakistan, Iran, and North Korea could increase the availability of nuclear materials and technology on the international marketplace, threatening the nuclear nonproliferation regime.[10] Alex Montgomery has countered that without the tacit knowledge that comes from deep experience with a nuclear weapons production program, states that receive nuclear assistance will still struggle to acquire nuclear weapons.[11] Yet these scholars do not systematically examine the effect of nuclear assistance on the spread of nuclear weapons.

Importing the Bomb

A state's ability to produce nuclear weapons often hinges on the availability of external assistance from a more advanced nuclear state. There are a number of common hurdles that states face as they attempt to develop a nuclear weapons program, but sensitive nuclear assistance from a more advanced nuclear power can help a state to overcome these technical and strategic challenges.

First, the designs for many sensitive nuclear technologies, such as uranium enrichment plants and implosion-type nuclear weapons, are not available in the public realm. States pursuing these technologies without external assistance must fashion designs for these complicated and advanced technologies indigenously. Second, the construction and successful operation of nuclear facilities requires much trial and error. Previous scholarship has emphasized the importance of

---

8. Singh and Way, "Correlates of Nuclear Proliferation," p. 862.
9. For recent quantitative studies of nuclear proliferation, see Gartzke and Kroenig, "Strategic Approach to Nuclear Proliferation."
10. Braun and Chyba, "Proliferation Rings"; and Chestnut, "Illicit Activity and Proliferation."
11. Alexander H. Montgomery, "Ringing in Proliferation," *International Security,* Vol. 30, No. 2 (Fall 2005), pp. 153–187.

tacit knowledge in creating successful nuclear weapons programs.[12] For example, the successful operation of a gaseous-centrifuge uranium enrichment plant requires the spinning of large metal cylinders at a rate of 300 meters per second. This is roughly the speed of sound. Inexperienced engineers often struggle to prevent the cylinders from spinning out of control and crashing on the ground. The kind of trial and error required for the indigenous development of advanced nuclear technology often ends in failure. For example, from 1981 to 1991, Iraq failed in multiple attempts to produce HEU, using several different methods including gaseous centrifuge, chemical enrichment, ion exchange, and laser isotope separation before finally settling on electromagnetic isotope separation.[13] Third, the development of a nuclear weapons infrastructure from scratch is an expensive enterprise. A state must, at a minimum, procure the relevant raw materials and technologies at home or on the open market; develop an advanced industrial and nuclear infrastructure; train, and provide for, a specialized cadre of physicists, mathematicians, engineers, and metallurgists; and provide adequate finances to continue to develop and support the program throughout its lifetime. For example, it is estimated that Iraq spent many billions of dollars in its unsuccessful bid to develop nuclear weapons. Fourth, states striving for a nuclear weapons capability must overcome these significant technical challenges under intense international pressure. Other states, international organizations, and nongovernmental organizations opposed to nuclear proliferation apply a variety of economic, diplomatic, and military pressures to dissuade states from fulfilling their nuclear ambitions. In 1981, for example, Iraq's nuclear reactor at Osiraq, then the centerpiece of Iraq's nuclear program, was destroyed by Israel in a preventive military strike.

Sensitive nuclear assistance can ease each of the challenges faced by potential nuclear weapon states. First, nuclear suppliers can provide the aspiring nuclear weapon state with proven designs for nuclear technology. With a guaranteed design in hand, scientists and technicians can leapfrog technical design stages and focus their effort on replicating a model that has proven effective elsewhere. For example, without access to Chinese nuclear bomb designs, it is believed that Pakistan would have had great difficulty developing a design for the implosion-type nuclear weapons that now constitute its nuclear arsenal.[14] Second, nuclear assistance can reduce the amount of trial and error needed to successfully operate

---

12. On the importance of tacit knowledge in nuclear weapon production programs, see Donald MacKenzie and Graham Spinardi, "Tacit Knowledge, Weapons Design, and the Uninvention of Nuclear Weapons," *American Journal of Sociology*, Vol. 100 (1995), pp. 44–99.
13. On Iraq's nuclear program, see, for example, Nuclear Threat Initiative, *Country Profiles*.
14. Corera, *Shopping for Bombs*, p. 46.

nuclear facilities. States supplying nuclear assistance can construct and even operate nuclear facilities for the recipient state. For example, when China provided Pakistan with uranium enrichment technology in the early 1980s, Chinese technicians remained in Pakistan until the uranium enrichment facility was fully operational.[15] In this way, the nuclear recipient benefits from the tacit knowledge acquired by the scientific community in the more advanced nuclear state. Third, importing sensitive nuclear technology can help states to economize on the costs of nuclear development. Procuring sensitive nuclear assistance from abroad can be less expensive than the indigenous development of a complete nuclear infrastructure. In fact, as we saw in previous chapters, states have often received substantial amounts of sensitive nuclear materials and technology at little or no cost because nuclear suppliers had a strategic interest in helping them acquire sensitive nuclear technology. As we saw in chapter 4, from 1958 to 1960, the Soviet Union provided China with the key component parts for the Lanzhou uranium enrichment facility and the Jiuquan plutonium reprocessing plants, partly because Moscow wanted to enhance China's defensive and deterrent capabilities against a possible U.S. attack. Fourth, and finally, sensitive nuclear assistance can help a state avoid international scrutiny. The receipt of sensitive nuclear materials and technology from abroad can quickly remake a state without a nuclear weapons program into a state with a latent nuclear weapons capability, presenting the international community with a fait accompli and preempting international efforts at dissuasion. As we saw in chapter 3, France provided Israel with sensitive nuclear assistance from 1959 to 1965, transforming Israel from a state with a rudimentary, civilian nuclear research program into a nuclear weapon state in less than a decade. The United States was strictly opposed to nuclear proliferation in Israel, but by the time U.S. intelligence agencies recognized the extent of Israel's nuclear program, the United States had few remaining policy options to dissuade Israel from its nuclear path.

The above discussion suggests that sensitive nuclear assistance is a contributing cause of nuclear proliferation. If this is true, we may expect a particularly tight relationship between sensitive nuclear transfers and nuclear proliferation. This logic leads us to the next hypothesis:

> Hypothesis 4: States that receive sensitive nuclear assistance will be more likely to acquire nuclear weapons.

It is possible, however, that other less sensitive types of nuclear assistance could also contribute to nuclear proliferation.[16] States that have a strong desire

---

15. Jones et al., *Tracking Nuclear Proliferation*, pp. 50, 57n.
16. The data section below contains definitions and more detailed information on the distinction between nonsensitive and sensitive nuclear assistance.

for nuclear weapons could solicit nonsensitive nuclear assistance, such as help in the construction of a nuclear research or power reactor, to reap the benefits of international nuclear cooperation in the development of a basic civilian nuclear infrastructure. Nonsensitive nuclear imports could free up domestic resources and create a cadre of specialized nuclear personnel that could later be channeled into the development of a militarized nuclear program. Moreover, nonsensitive nuclear assistance could also lead to the creation of a domestic nuclear complex that becomes a political constituency within internal political debates, pushing for further nuclear development that could eventually result in weaponization.

A relationship between nonsensitive nuclear assistance and nuclear proliferation has been suspected by academics and nuclear nonproliferation professionals. For example, Matthew Furhmann writes: "The history of nuclear proliferation reveals that civilian nuclear energy can aid nuclear weapons production."[17] Peter Lavoy agrees that nonsensitive nuclear cooperation "dramatically reduced the costs of undertaking serious nuclear research and development for dozens of nations around the world" and that "in a handful of cases, highly determined governments succeeded in producing nuclear weapons from so-called peaceful nuclear technologies."[18] Similarly, Leonard Weiss finds that the nonsensitive nuclear assistance that transpired under the U.S.-sponsored Atoms for Peace program contributed to the international spread of nuclear weapons: "Did the 50-year-old Atoms for Peace program accelerate nuclear weapons proliferation? The jury has been in for some time on this question, and the answer is yes."[19] From this line of analysis, we may expect nonsensitive nuclear assistance also to lead to nuclear proliferation. This logic leads to the next hypothesis:

> Hypothesis 5: States that receive nonsensitive nuclear assistance will be more likely to acquire nuclear weapons.

There are clearly other explanations for why states acquire nuclear weapons. I therefore control for a wide set of opportunity and willingness determinants of nuclear proliferation. I discuss these variables in the below sections as well as describe the data and examine the evidence for the above hypotheses.

Empirical Analysis

To examine the relationship between sensitive nuclear assistance and nuclear proliferation, I employ qualitative and quantitative research methods. Nuclear proliferation and sensitive nuclear assistance are both rare events. From 1945 to

---
17. Matthew Fuhrmann, "Taking a Walk on the Supply Side."
18. Peter R. Lavoy, "The Enduring Effects of Atoms for Peace," *Arms Control Today* (December 2003).
19. Leonard Weiss, "Atoms for Peace," *Bulletin of Atomic Scientists*, Vol. 59, No. 6 (November 2003).

2000, the time period under study, nine countries acquired nuclear weapons, of which three (Israel, China, and Pakistan) received sensitive nuclear assistance.[20] The relatively small number of positive cases allows me to examine the role of sensitive nuclear assistance in the positive cases and to compare these countries to similar countries that did not receive sensitive nuclear assistance. The qualitative analysis is only the first step, however. To analyze the relationship between sensitive nuclear assistance and nuclear proliferation in the entire universe of cases and to control for potentially confounding factors, a large-N statistical analysis will form the core of the empirical investigation.

Case Studies

A brief review of important cases of nuclear proliferation demonstrates that assistance from abroad can be an important factor in determining whether or not a state eventually acquires nuclear weapons. As we saw in chapter 3, in the late 1950s, Israel's nuclear program consisted of nothing more than a national atomic energy commission and a small research reactor. From 1959 to 1965, however, France provided sensitive nuclear assistance to Israel, greatly enhancing Israel's ability to produce nuclear weapons. France constructed a large plutonium-producing nuclear reactor and a plutonium reprocessing facility at Dimona, transferred a nuclear weapon design, trained Israeli scientists at nuclear facilities in France, and allowed Israeli observers at French nuclear weapon tests. By 1967, after sustained French assistance, Israel was able to assemble its first nuclear weapon.

Many other states with nuclear arsenals received substantial assistance from abroad. In fact, much of the history of nuclear proliferation can be read as a chain of countries offering sensitive nuclear assistance. As we saw in chapter 4, from 1958 to 1960, the Soviet Union provided China with key component parts for uranium enrichment and plutonium reprocessing plants and trained Chinese technicians, contributing to China's ability to conduct its first nuclear weapon test in 1964. Chapter 4 demonstrated that China itself became a nuclear supplier. From 1981 to 1986, China transferred significant quantities of HEU, uranium enrichment technologies, and a nuclear weapon design to Pakistan. While Pakistan refrained from testing a nuclear device until 1998, it is believed that Chinese assistance enabled Pakistan to assemble its first nuclear weapon in 1992. Chapter 4 also showed us that Pakistan then dispersed sensitive nuclear technology and

---

20. North Korea received sensitive nuclear assistance in the 1990s and tested its first nuclear device in 2006. Because North Korea acquired nuclear weapons after the end of the time period under investigation, it is not included in the analysis as a positive case of nuclear proliferation.

materials to Iran, Libya, and North Korea from 1987 to 2002. This assistance also contributed to nuclear proliferation. In 2006, North Korea tested its first nuclear weapon, and many estimate that Iran could be capable of producing a nuclear device in a few short years. A 2007 U.S. National Intelligence Estimate predicted that Iran would be able to produce nuclear weapons as early as 2009, thanks in large part to uranium enrichment assistance from Pakistan.[21]

On the other hand, states with a persistent demand for nuclear weapons that did not acquire substantial international assistance failed to sustain national nuclear weapons programs. Egypt, over the course of many decades, has been rebuffed in numerous attempts to secure an international nuclear supplier; to this day, Egypt lacks a nuclear weapons arsenal. Beginning in the 1960s, Egypt sought sensitive nuclear assistance from the Soviet Union, China, and France but was denied by all three states.[22] There is also evidence to suggest that Egyptian officials met with representatives from the A. Q. Khan network, but as far as we know, Egypt never received sensitive nuclear assistance from Pakistan.[23] Unlike Israel and other current nuclear weapon states that received substantial imports of sensitive nuclear materials and technology, Egypt currently maintains a rudimentary civilian nuclear program. There are other states that have shown a historical interest in nuclear weapons but have not yet acquired the bomb. This group includes states that have received scant sensitive nuclear assistance from abroad, including Iraq and Taiwan, and states that have not received any sensitive nuclear assistance, such as Argentina, Saudi Arabia, Syria, and South Korea.[24]

Nuclear Proliferation Data

To test the effect of international nuclear assistance on the spread of nuclear weapons, I perform a quantitative analysis on all states in the international system from 1945 to 2000.[25] To conduct the analysis, I draw on the original nuclear assistance dataset presented in chapter 2.

---

21. Office of the Director of National Intelligence, *Iran: Nuclear Intentions and Capabilities*, November 2007.
22. On Egypt's quest to acquire sensitive nuclear materials and technology, see Nuclear Threat Initiative, *Country Profiles;* and Ariel Levite and Emily B. Landau, "Arab Perceptions of Israel's Nuclear Posture, 1960–1967," *Israel Studies*, Vol. 1, No. 1 (Spring 1996), pp. 34–59.
23. On possible Egyptian connections with the A. Q. Khan network, see, for example, "The A. Q. Khan Network: Case Closed?" Hearing before the Subcommittee on International Terrorism and Nonproliferation of the Committee on International Relations. House of Representatives, May 25, 2006.
24. Syria recently received assistance from North Korea on the construction of a nuclear reactor, but there is no evidence to indicate that Syria received any sensitive nuclear assistance. For more information on this case, see appendix D.
25. The unit of analysis is the country-year.

Table 5.1. Nuclear weapons proliferation, 1945–present

COUNTRY	DATE
United States	1945
USSR/Russia	1949
United Kingdom	1952
France	1960
China	1964
Israel	1967
India	1988
South Africa	1982–1990
Pakistan	1990
Belarus	Never
Kazakhstan	Never
Uzbekistan	Never
North Korea	Never

Notes: Israel began separating plutonium in 1966 and assembled two makeshift nuclear weapons on the eve of the 1967 war. Cohen, *Israel and the Bomb,* pp. 273–276. India first readied two dozen nuclear weapons for quick assembly and delivery by aircraft between 1988 and 1990. Perkovich, *India's Nuclear Bomb,* pp. 293–297. South Africa constructed its first nuclear device in 1979, but the first deliverable device wasn't ready until 1982. The 1982 device was deliverable only in the sense that it could have been "kicked out of the back of a plane." South Africa began dismantling its nuclear program in 1990. Albright, "South Africa's Secret Nuclear Weapons," p. 43. Pakistan had enough enriched uranium to produce nuclear weapons in 1987, but it was not until 1990 that it manufactured the metal components for a nuclear device. Jones et al., *Tracking Nuclear Proliferation,* pp. 132, 140n. The nuclear weapons in Belarus, Ukraine, and Kazakhstan were generally believed to have been under de facto Russian/CIS control and that in order to acquire the nuclear weapons, the newly independent states would have had to seize them from Russian forces. See, for instance, Steven E. Miller, "The Case against a Ukrainian Nuclear Deterrent," *Foreign Affairs,* Vol. 73, No. 3 (Summer 1993), pp. 67–80. North Korea tested a first nuclear device in October 2006 and a second device in May 2009. Many experts believe, however, that these tests were failures and evidence of the fact that North Korea still lacks a deliverable nuclear weapon. See, for instance, David Sanger, "Tested Early by North Korea, Obama has Few Options," *New York Times,* May 29, 2009.

The dichotomous dependent variable is *Acquire.*[26] This variable measures whether a state acquires nuclear weapons in a given year.[27] A state is coded as acquiring nuclear weapons when it first explodes a nuclear device or, if it does not conduct a nuclear test, when it first assembles a deliverable nuclear weapon. Nuclear weapon states are listed in table 5.1.

---

26. Data on nuclear acquisition are drawn from Gartzke and Kroenig, "Strategic Approach to Nuclear Proliferation."

27. I do not include variables measuring state decisions to explore, pursue, or possess nuclear weapons because my theoretical interest is on the effect of nuclear assistance on nuclear acquisition. Nevertheless, robustness tests performed using these alternate measures of nuclear proliferation produce similar results.

Table 5.2. Recipients of sensitive nuclear assistance

RECIPIENT	YEAR OF FIRST ASSISTANCE	SUPPLIER(S)	TYPE OF ASSISTANCE
China	1958	USSR	Plutonium reprocessing, uranium enrichment
Israel	1959	France	Plutonium reprocessing, nuclear weapon design
Japan	1971	France	Plutonium reprocessing
Pakistan	1974	France, China	Plutonium reprocessing, uranium enrichment, nuclear weapon design
Taiwan	1975	France	Plutonium reprocessing
Iraq	1976	Italy	Plutonium reprocessing
Brazil	1979	Germany	Plutonium reprocessing, uranium enrichment
Egypt	1980	France	Plutonium reprocessing
Iran	1984	China, Pakistan	Uranium enrichment, nuclear weapon design
Algeria	1986	China	Plutonium reprocessing
Libya	1997	Pakistan	Uranium enrichment, nuclear weapon design
North Korea	1997	Pakistan	Uranium enrichment, nuclear weapon design

I also construct variables to examine the relationship between various kinds of nuclear assistance and nuclear proliferation. *Sensitive nuclear assistance* measures whether a state has ever received the key materials and technologies necessary for the construction of a nuclear weapons arsenal from a capable nuclear-supplier state. A full definition of sensitive nuclear assistance, along with a discussion of operationalization, measurement, and coding, can be found in chapters 1 and 2. A list of countries that have received sensitive nuclear assistance can be found in table 5.2.[28]

It is possible, however, that sensitive nuclear assistance is endogenous to nuclear acquisition. In other words, nuclear supplier states may channel sensitive nuclear assistance to states that they believe are most likely to develop nuclear weapons.[29] If this is the case, any relationship between sensitive nuclear assistance and nuclear acquisition may be because states that are likely to acquire

---

28. See appendices C, D, and E for a description of the cases of sensitive nuclear assistance, a list and description of selected cases that do not qualify as sensitive nuclear assistance according to the above definition, an explanation of key coding decisions, and the sources used in coding the variable.

29. Endogenity between nuclear assistance and nuclear acquisition hinges on the motivations of the nuclear suppliers. I have shown that suppliers provide nuclear assistance based on strategic considerations. One question that is raised, however, is why states are willing to provide sensitive nuclear assistance but have never exported complete nuclear arsenals. Endogeneity may be present if states are unwilling to provide complete nuclear arsenals because they believe the recipient will acquire nuclear weapons absent nuclear assistance but decide to provide some assistance anyway to gain leverage over the recipient or to gain intelligence on the recipient's nuclear program. This theoretical

nuclear weapons are more likely to receive sensitive nuclear assistance. To test for this endogeneity bias, I construct *Pr. (Sensitive nuclear assistance)*. This variable is the predicted probability that a state will receive sensitive nuclear assistance based on the statistical model developed in chapter 2. States in threatening security environments, that are relatively closed to the international economy, and that are not closely aligned with a superpower patron are more likely to receive sensitive nuclear assistance.[30]

To test the hypothesis that nonsensitive nuclear transfers contribute to nuclear proliferation, I construct *Nonsensitive nuclear assistance*. This dichotomous variable measures whether a state has ever received nonsensitive nuclear assistance from a capable nuclear supplier. The first significant step in the development of a nuclear program is the construction of a nuclear reactor. A state is coded as having received nonsensitive nuclear assistance from the first year in which a foreign country begins the construction of a nuclear research reactor or a nuclear power reactor on its soil.[31] States that have never received assistance in the construction of a nuclear reactor are coded as having never received nonsensitive nuclear assistance.

It is possible that a measure of nonsensitive nuclear assistance focusing narrowly on reactors could miss some important transfers. Hypothetically, a state with a nuclear program could develop all of its nuclear reactors indigenously but receive other types of nonsensitive nuclear assistance from abroad. In reality, however, there is little cause for concern. Very few states with a nuclear program developed all of their nuclear reactors indigenously. In fact, nearly every state with a nuclear program received its first nuclear reactor from either the United States or the Soviet Union during the Cold War. This measure also has the added benefit of excluding less salient types of nuclear assistance, such as scientific exchanges and assistance in the surveying and mining of natural uranium, to countries that lack a basic nuclear infrastructure.

I also include a number of variables to control for the factors thought to influence the likelihood of nuclear acquisition. For more detail on each, see the

---

concern does not appear to pose a problem, however. The findings of previous chapters suggest that suppliers assume that assistance is critical to whether the recipient acquires nuclear weapons, mitigating concerns about endogenity. Moreover, some of the evidence presented in chapter 3 could suggest that states provide nuclear assistance to promote targeted nuclear proliferation but stop short of transferring complete nuclear arsenals in order to avoid international condemnation.

30. I also generated alternate *Pr. (sensitive nuclear assistance)* variables based on other models that included additional variables that could potentially affect the probability of sensitive nuclear assistance, including level of economic development and industrial capacity. The use of these alternate variables did not affect the results.

31. Data on the construction of nuclear reactors are drawn from the IAEA's list of nuclear research reactors in the world and the World Nuclear Association's list of nuclear power reactors. World Nuclear Association, "Nuclear Power Reactors."

discussion in appendix B. To assess a country's domestic capability to produce nuclear weapons, I include a measure of GDP per capita and a separate measure that gauges a country's industrial capacity. To test the effect of a state's security environment on its risk of acquiring nuclear weapons, I code variables measuring whether a country is involved in at least one enduring rivalry; an average of the number of militarized interstate disputes (MIDs) per year in which a country is involved; whether the country enjoys the protection of a nuclear umbrella from any nuclear-armed state; and whether the country is in a defense pact with a superpower. To gauge institutional and economic determinants of nuclear acquisition, I code variables that measure domestic regime type; changes in domestic regime type over time; the percentage of democracies in the international system; dependence on international trade; and changes in a state's dependence on international trade over time.

Data Analysis

I employ event-history models to test claims about the correlates of nuclear acquisition.[32] Event-history models, also known as hazard, survival, or duration models, are widely used statistical estimators designed to calculate the risk that any given observation will experience a specific event. Hazard models are frequently used, for example, in epidemiology to determine the risk that an individual will contract cancer. In this study, I employ hazard models to calculate the risk that a country will acquire nuclear weapons.[33]

Several types of statistical analysis prove useful in exploring the evidence for or against each of the hypotheses described earlier. To begin the investigation, I replicate the findings of previously published research on nuclear proliferation by Singh and Way. These findings are presented in table 5.3, model 1.[34] After replicating Singh and Way's findings, I add the nuclear assistance variables to test the

---

32. Janet M. Box-Steffensmeier and Bradford S. Jones, "Time Is of the Essence," *American Journal of Political Science*, Vol. 41 (1997), pp. 1414–1461.

33. Following Singh and Way, "Correlates of Nuclear Proliferation," I estimate parametric discrete-time hazard models using a Weibull distribution to characterize the baseline hazard function. Using Cox-proportional hazard, logit, or probit estimators produces virtually identical results. Robust standard errors are adjusted for clustering by country.

34. Using the data and the techniques described in Singh and Way, "Correlates of Nuclear Proliferation," I was unable to replicate the reported findings exactly. The reason for the differences between their reported results and my analysis remain unaccounted for. My findings, while not identical, are quite similar. The signs on the coefficients of the variables and their level of statistical significance are the same with a single exception. In contrast to Singh and Way, I find that the democracy variable attains statistical significance in the model estimating nuclear acquisition. Singh and Way report 149 countries in their results. My analysis includes 156 countries. Singh and Way report 5,784 observations. My analysis is conducted on 5,822 observations.

effect of nuclear assistance on nuclear proliferation. These findings are presented in table 5.3, models 2–4. I also control for other factors, NPT membership and superpower dependence, that were omitted from Singh and Way's study. These findings are reported in table 5.3, Models 5–7.

The replication analysis is only the first step, however. It is possible that many states never acquire nuclear weapons because they never seriously considered the nuclear option. By including states that never possess a nuclear weapons production program in the analysis, I may be overrepresenting the relationship between the receipt of nuclear assistance and nuclear acquisition, biasing the results. To assess the relationship between nuclear assistance and nuclear acquisition among only those states that desire nuclear weapons, I use a censored hazard model of the risk of nuclear acquisition contingent on a state possessing a nuclear weapons production program as measured by Jo and Gartzke.[35] These findings are presented in table 5.4, models 1–3. I also estimate a censored hazard model of the risk of nuclear acquisition contingent on a state exploring the nuclear option as measured by Singh and Way.[36] These findings are presented in table 5.4, models 4–6. Finally, I perform a number of statistical tests to examine the robustness of these findings. The robustness checks are reported in appendix B.

I first examine the hypothesis that sensitive nuclear assistance is positively related to nuclear acquisition. Hypothesis 4 posits that states that receive sensitive nuclear assistance will be more likely to acquire nuclear weapons. Turning first to the hazard models, we see that the relationship between sensitive nuclear assistance and nuclear acquisition is positive and statistically significant in each and every model. Next, an examination of the censored hazard models reveals a similar pattern. Again, the sign on the coefficient is positive and statistically significant in each model. There is strong empirical support for the causal significance of sensitive nuclear assistance for understanding nuclear proliferation.

Next, I examine the possibility that the relationship between sensitive nuclear assistance is endogenous to nuclear acquisition. It may be that states that are likely to acquire nuclear weapons are more likely to receive sensitive nuclear assistance. If this is the case, we should expect a positive relationship between the probability that a state will receive sensitive nuclear assistance and nuclear acquisition. Turning first to the hazard models, we see that the sign on the coefficient of Pr. (*Sensitive nuclear assistance*) is positive but not statistically significant. Neither is this variable statistically significant in the censored hazard models. Moreover, in the censored hazard models, the sign on the coefficient is inconsistent, switching from negative to positive depending on model specification. There is

---

35. Jo and Gartzke, "Determinants of Nuclear Weapons Proliferation."
36. Singh and Way, "Correlates of Nuclear Proliferation."

Table 5.3. Hazard models of nuclear proliferation, 1945–2000

INDEPENDENT VARIABLE	1	2	3	4	5	6	7
Sensitive nuclear assistance		1.769** (0.738)			1.826** (0.718)		
Pr. (Sensitive nuclear assistance)			1.390 (1.811)			1.898 (1.561)	
Nonsensitive nuclear assistance				−0.758 (0.873)			−0.707 (0.738)
GDP per Capita	0.0003 (0.0003)	0.0003 (0.0003)	0.0002 (0.0003)	0.0004 (0.0003)	0.0006** (0.0003)	0.0002 (0.0002)	0.0005** (0.0002)
GDP squared	−2.38e-08* (1.44e-08)	−2.61e-08** (1.27e-08)	−1.97e-08 (1.29e-08)	−2.49e-08 (1.64e-08)	−4.00e-08** (1.70e-08)	−6.42e-09 (1.04e-08)	−2.96e-08*** (1.14e-08)
Industrial capacity index	3.106**** (0.752)	3.548**** (0.561)	3.223**** (0.492)	3.400*** (1.224)	4.059**** (1.053)	4.789*** (1.768)	3.494**** (0.784)
Enduring rivalry	2.219* (1.194)	1.980* (1.154)	2.139 (1.353)	2.287** (1.164)	1.865 (1.187)	1.298 (2.063)	2.187** (1.116)
Dispute involvement	0.242** (0.119)	0.216** (0.102)	0.229 (0.234)	0.232* (0.135)	0.222** (0.105)	0.516 (0.452)	0.188 (0.121)
Alliance	−1.166 (0.870)	−1.098 (0.892)	−0.872 (0.881)	−1.237 (0.831)			
Superpower dependent					−2.233* (1.155)	−2.192* (1.281)	−2.264** (0.887)
Democracy	0.083* (0.047)	0.111* (0.061)	0.107 (0.084)	0.084* (0.049)	0.107 (0.073)	0.056 (0.150)	0.083 (0.064)

(continued)

Table 5.3. Hazard models of nuclear proliferation, 1945–2000 (Continued)

INDEPENDENT VARIABLE	1	2	3	4	5	6	7
Democratization	0.026	0.007	0.006	0.038	-0.051	-0.051	0.024
	(0.118)	(0.111)	(0.109)	(0.115)	(0.104)	(0.098)	(0.087)
Percentage of	-0.105	-0.119	-0.048	-0.127	-0.020	0.144*	-0.016
democracies	(0.101)	(0.089)	(0.078)	(0.101)	(0.101)	(0.076)	(0.121)
NPT					-18.012****	-21.523****	-18.787****
					(0.906)	(2.544)	(1.008)
Economic openness	-0.005	-0.004	0.0009	-0.007	-0.017	-0.009	-0.012
	(0.017)	(0.021)	(0.020)	(0.020)	(0.023)	(0.030)	(0.023)
Economic liberalization	0.0009	-0.002	-0.002	0.006	-0.002	0.009	0.020
	(0.020)	(0.024)	(0.028)	(0.023)	(0.026)	(0.041)	(0.029)
Constant	-6.847**	-6.424**	-8.762****	-6.624**	-9.932**	-16.114****	-9.655**
	(3.097)	(2.950)	(2.413)	(3.209)	(4.210)	(3.774)	(4.593)
Ancillary parameter (p)	0.886	0.747	0.899	1.059	0.884	1.376	1.164
Standard error (p)	0.207	0.120	0.191	0.279	0.453	0.398	0.689
Log likelihood	-21.405	-19.549	-18.304	-21.108	-13.911	-9.380	-15.280
Number of countries	156	156	154	156	156	154	156
Total observations	5822	5822	5493	5822	5822	5493	5822

Note: Statistically significant parameter estimators are denoted by *p < .1; **p < .05; ***p < .01; ****p < .001. Coefficients are estimates for parametric survival models with a Weibull distribution; robust standard errors, adjusted for clustering by country, are in parentheses. GDP = gross domestic product. Pr. (*Sensitive nuclear assistance*) is calculated using the Affinity of Nations Index, which relies on data of UN General Assembly voting. Gartzke, *Codebook for the Affinity of Nations Index*. In chapter 2 we found that vulnerability to superpower pressure as measured by voting patters in the UN General Assembly is an important predictor of which states receive sensitive nuclear assistance. Models that include this measure lose observations for countries that were not members of the UN. Using an alternate estimate for the probability that a state will receive sensitive nuclear assistance that draws on an alternative measure of superpower dependence and that does not lose any observations produces virtually identical results.

Table 5.4. Censored hazard models of nuclear proliferation, 1945–2000

INDEPENDENT VARIABLE	1	2	3	4	5	6
Sensitive nuclear assistance	1.320* (0.781)			2.116** (1.009)		
Pr. (Sensitive nuclear assistance)		−0.155 (1.742)			1.588 (1.739)	
Nonsensitive nuclear assistance			−6.749* (3.690)			−2.605** (1.077)
GDP per Capita	0.0007 (0.0005)	−0.0005 (0.0008)	0.0003 (0.0006)	0.001** (0.0005)	0.0005 (0.0007)	0.0007* (0.0004)
GDP squared	−4.47e-08 (2.92e-08)	3.91e-08 (6.31e-08)	−3.59e-08 (4.88e-08)	−5.12e-08** (2.12e-08)	−1.29e-08 (4.07e-08)	−3.33e-08 (2.55e-08)
Industrial capacity index	3.166** (1.396)	2.584 (2.311)	3.872 (2.763)	20.540**** (2.282)	21.785**** (1.955)	21.414**** (2.484)
Enduring rivalry	1.655 (2.002)	1.253 (3.260)	3.387* (1.919)	2.795* (1.506)	2.297 (2.393)	3.278* (1.787)
Dispute involvement	0.022 (0.239)	0.334 (0.382)	0.326 (0.327)	0.280** (0.128)	0.519* (0.308)	0.366*** (0.131)
Superpower dependent	−3.209 (2.982)	−1.446 (2.625)	−1.890 (1.435)	−2.474 (2.220)	−2.101 (3.055)	−1.735 (1.149)
Democracy	0.069 (0.079)	−0.016 (0.179)	−0.014 (0.129)	0.188** (0.092)	0.097 (0.145)	0.125** (0.060)

(continued)

Table 5.4. Censored hazard models of nuclear proliferation, 1945–2000 (Continued)

INDEPENDENT VARIABLE	MODEL					
	1	2	3	4	5	6
Democratization	-0.105	-0.080	-0.011	-0.078	-0.026	-0.020
	(0.092)	(0.096)	(0.136)	(0.051)	(0.072)	(0.055)
Percentage of	0.144	0.286***	-0.128	0.092	0.246**	-0.029
democracies	(0.176)	(0.106)	(0.289)	(0.115)	(0.111)	(0.204)
NPT	-20.334****	-29.788*	-20.005****	-19.788****	-23.124****	-19.971****
	(2.480)	(16.771)	(3.030)	(1.634)	(1.981)	(1.490)
Economic openness	-0.035	0.040	-0.005	-0.089*	-0.051	-0.055*
	(0.052)	(0.070)	(0.051)	(0.050)	(0.063)	(0.032)
Economic liberalization	0.058	-0.008	0.066	-0.027	-0.059	-0.015
	(0.105)	(0.191)	(0.162)	(0.050)	(0.044)	(0.062)
Constant	-13.962**	-16.963***	-13.534	-33.349****	-39.134****	-31.984****
	(6.114)	(5.827)	(8.980)	(6.761)	(6.849)	(7.199)
Ancillary parameter (p)	1.496	1.565	4.458	1.623	1.939	2.512
Standard error (p)	0.635	0.643	1.751	0.503	0.586	0.976
Log likelihood	-9.043	-4.710	-4.527	-11.352	-7.351	-11.307
Number of countries	18	17	18	23	20	23
Total observations	386	311	386	461	353	461

Note: Statistically significant parameter estimators are denoted by *p < .1; **p < .05; ***p < .01; ****p < .001. Coefficients are estimates for parametric survival models with a Weibull distribution; robust standard errors, adjusted for clustering by country, are in parentheses. GDP = gross domestic product.

no support for the idea that nuclear acquisition is endogenous to sensitive nuclear assistance.[37] This suggests that sensitive nuclear assistance is not merely epiphenomenal to nuclear proliferation. In other words, states that receive sensitive nuclear assistance are more likely to acquire nuclear weapons, not vice versa.

The next hypothesis focuses on the effect of nonsensitive nuclear assistance on nuclear acquisition. States that receive assistance in the nonsensitive nuclear realm may also be more likely to acquire nuclear weapons. As a reminder, we should expect a positive relationship if nonsensitive nuclear assistance is also a contributing cause of nuclear proliferation. Turning first to the hazard models, we do not find support for this hypothesis. The sign on the coefficient of the nonsensitive nuclear assistance variable is in the unexpected direction and is not statistically significant. The censored hazard models provide no further support for this hypothesis. In the censored hazard models, nonsensitive nuclear assistance is statistically significant, but the signs on the coefficients are negative.[38] This suggests, contrary to theoretical expectation, that among states with a nuclear weapons production program, those that receive nonsensitive nuclear assistance are actually less likely to acquire nuclear weapons. This result may suggest that one of the grand bargains of the NPT was successful and that states willingly traded their right to develop nuclear weapons in exchange for nonsensitive nuclear technology.

Next, I examine the control variables to assess the relative support for supply-side, as opposed to demand-side, explanations of nuclear proliferation. International assistance and domestic capacity are the primary means by which a state can acquire the capability to produce nuclear weapons. I have already found a relationship between sensitive nuclear assistance and nuclear proliferation. *Industrial capacity index* is positive in every model and is statistically significant in eleven of the thirteen models. States above a certain level of industrial development are more likely to acquire nuclear weapons. Taken together with the findings about the relationship between sensitive nuclear assistance and nuclear proliferation, we find strong support for the supply-side approach to understanding nuclear proliferation. States that can acquire nuclear weapons, either through international assistance or domestic capacity, are more likely to do so.

Turning to the demand-side variables, we find that the *Rivalry* and *Disputes* variables are positive in every model and are statistically significant in roughly half of the models. This finding suggests that states in a threatening security

---

37. Pr. (*sensitive nuclear assistance*) is not included in the models with *Sensitive nuclear assistance* due to concerns about multicollinearity. The bivariate correlation between these variables is r = .84.

38. I also tried a level of nuclear assistance variable coded: 0 = no assistance, 1 = nonsensitive assistance, 2 = sensitive assistance. This variable was not statistically significant in any of the models.

environment are more likely to acquire nuclear weapons. There is no discernable relationship between a security guarantee from a nuclear power and nuclear acquisition. *Alliance* is not statistically significant in any of the models in which it is included. The alternate *Superpower dependence* variable, however, is negative in every model in which it is included and is statistically significant in the hazard models. Consistent with the logic that motivates Hypothesis 3, states that are vulnerable to superpower pressure are less likely to engage in activities related to nuclear proliferation. The other variables representing state demand for nuclear weapons are not statistically significant across a substantial proportion of the models. Taken together, I find modest support for the demand-side approach to the study of nuclear proliferation.

We next turn to *NPT*, which does not fit neatly into either the opportunity or willingness categories. The relationship between *NPT* and nuclear acquisition is negative and significant in each and every model in which it is included. States that are members of the NPT are less likely to acquire nuclear weapons. It may be that states that are members of the NPT have less opportunity to develop nuclear weapons because their nuclear programs are subject to international safeguards. It may also be that NPT members are less willing to pursue nuclear weapons because they are satisfied with the benefits of NPT membership, or because a visible international commitment to the nonproliferation regime undermines domestic political factions that favor nuclear proliferation. It does not appear, however, that *NPT* is merely a signal of intent that stands for other willingness variables. The effects of the other willingness variables are not greatly attenuated by the inclusion of *NPT*.

We have seen that sensitive nuclear assistance has a statistically significant effect on nuclear proliferation. How large is this effect substantively? Table 5.5 interprets the substantive effect of these variables on nuclear acquisition, using the results from the hazard model reported in table 5.3, model 5, and the censored hazard model reported in table 5.4, model 1. The entries represent the percent change in the baseline hazard ratios of nuclear acquisition for a given change in the independent variable. Focusing my comments on the results from the censored hazard model, the table reveals that a state that receives sensitive nuclear assistance has a risk of acquiring nuclear weapons that is about three times larger than a similar state that does not receive sensitive nuclear assistance. Sensitive nuclear assistance does not just have a statistically significant effect but also a substantively significant effect on nuclear acquisition.

Turning now to the substantive effects of the control variables, table 5.5 shows that industrial capacity has a substantive impact on nuclear acquisition. States above a certain threshold of industrial capacity have a hazard ratio of acquiring nuclear weapons that is over 22 times larger than similar states below the

Table 5.5. Substantive effects of variables on the likelihood of nuclear proliferation, 1945–2000

	PERCENT CHANGE IN THE HAZARD RATIOS	
VARIABLE	NONCENSORED	CENSORED
Sensitive nuclear assistance	+521	+274
Industrial capacity threshold	+5,693	+2,270
Increase in frequency of MIDs (two more/year)	+50	+04
Rivalry	+545	+423
NPT membership	−100	−100

Note: Hazard ratios are based on the models reported in table 5.3, model 5 and table 5.4, model 1.
GDP = gross domestic product; MID = militarized interstate dispute.

industrial capacity threshold. NPT membership also has a noticeable impact on the patterns of sensitive nuclear assistance. States that are members of the NPT are about half as likely to acquire nuclear weapons as similar states that belong to the NPT. States that are involved in an enduring rivalry are over four times more likely to acquire nuclear weapons than are similar states not so involved. In contrast, MIDs appear to have a substantively trivial impact on nuclear acquisition. Increasing the number of MIDs in which a state is involved by two per year increases the risk that it will acquire nuclear weapons by only 4 percent.

Conclusion

This chapter sought to explain why states acquire nuclear weapons. I found that in order to explain patterns of nuclear proliferation, one must look to international transfer of sensitive nuclear materials and technology. States that receive sensitive nuclear assistance are more likely to acquire nuclear weapons than are similar states that do not receive sensitive nuclear assistance. The receipt of sensitive nuclear assistance helps potential nuclear proliferators overcome the common obstacles that states encounter as they attempt to develop a nuclear arsenal. By importing the bomb, states can leapfrog technical design stages, benefit from tacit knowledge in more advanced scientific communities, economize on the cost of nuclear weapons development, and avoid international scrutiny.

Arguments that contend that nonsensitive nuclear assistance contributes to nuclear proliferation do not find support in the data. There is no discernable positive relationship between nonsensitive nuclear assistance and nuclear proliferation. This finding validates this book's focus on the specific problem of sensitive nuclear assistance.

The argument of this chapter began with the simple insight that the ability to construct nuclear weapons spreads from state to state. Scholarly approaches to nuclear proliferation have focused largely on the question of *why* states want nuclear weapons. The existing literature has paid less attention, however, to *how* states acquire nuclear weapons. Indeed, it is somewhat peculiar that studies about the "proliferation," "diffusion," and "spread" of nuclear weapons have not explicitly recognized that nuclear weapons technologies and materials literally spread from state to state. This chapter demonstrates that a complete explanation of nuclear proliferation requires an adequate understanding of both the supply of, and demand for, nuclear weapons.

In a Council on Foreign Relations special report on nuclear energy, Charles D. Ferguson writes that "humanity confronts two stark risks: catastrophes caused by climate change and annihilation by nuclear war."[39] He notes that these dual risks have created a genuine policy dilemma related to the spread of nuclear technology. The fear of global climate change has led many to promote nuclear energy, a technology that emits few greenhouse gases into the atmosphere, as a means to meet increasing global energy demands while addressing global climate change. Yet many nuclear nonproliferation professionals retort that the worldwide diffusion of nuclear technology for peaceful purposes can easily be hijacked by military nuclear programs and contribute to the international spread of nuclear weapons. The findings of this chapter suggest that this dilemma does not actually exist. The diffusion of civilian nuclear technologies does not present a significant proliferation risk, it is only the spread of sensitive nuclear technologies that directly contributes to the international spread of nuclear weapons. States concerned about nuclear proliferation must refrain from transferring sensitive nuclear materials and technology to nonnuclear weapon states. States can, however, provide nonsensitive civilian nuclear assistance to promote peaceful uses of nuclear energy and counter global climate change with very little risk of contributing to the international spread of nuclear weapons.

---

39. Charles D. Ferguson, *Nuclear Energy*, Council on Foreign Relations, Special Report 28 (2007).

CONCLUSION
Preventing Nuclear Proliferation

Why do states provide sensitive nuclear assistance to nonnuclear weapon states, contributing to the international spread of nuclear weapons? Why do other nuclear-capable states refrain from providing such assistance? Few questions are more important for international relations scholars interested in understanding the role of nuclear weapons in international politics and for policymakers interested in stopping the spread of nuclear weapons.

Yet scholars have not systematically addressed this topic. Media reports and policy analyses generally claim that sensitive nuclear exports are, like any other export, driven by economic considerations. States, especially states in dire economic straits, export sensitive nuclear materials and technology abroad in search of economic gain. These states so badly want to improve their economic situation that they overlook the threat that nuclear proliferation poses to international peace and security—as well as to themselves.

The findings of this book contest this common belief. States that provide sensitive nuclear assistance understand the security threat posed by nuclear proliferation. They know that it is worse for some countries than it is for others. The strategic theory of nuclear proliferation presented in this book is derived from a single basic insight: the spread of nuclear weapons is worse for powerful states than it is for weak states. Power-projecting states, states with the ability to project conventional military power over a particular target state, incur a number of strategic costs when that target state acquires nuclear weapons. On the other hand, non-power-projecting states, states that lack the ability to project power over a particular target state, do not incur as many strategic costs when that target state acquires nuclear weapons.

This is an argument about capabilities, not intentions. Power-projecting states incur significant strategic costs whether nuclear weapons spread to friends or foes. While the threat is most severe when nuclear weapons are acquired by enemy states, nuclear proliferation, even to friendly states, causes many problems for power-projecting states. On the other hand, non-power-projecting states have fewer constraints placed on their conventional military freedom of action as nuclear weapons spread. Again, this is true whether the bomb spreads to friendly or hostile states. For this reason, non-power-projecting states are, on average, less threatened by the spread of nuclear weapons.

These differential effects of nuclear proliferation on states with different levels of conventional military power lead to three conditions under which states provide sensitive nuclear assistance. First, the better able a state is to project power over a particular state, the less likely it is to provide sensitive nuclear assistance to that state. States avoid imposing strategic costs on themselves. Second, states can provide sensitive nuclear assistance to impose strategic costs on powerful rivals. For this reason, states are more likely to provide sensitive nuclear assistance to a state with which they share a common enemy. Third, states that are less vulnerable to superpower pressure are more likely to provide sensitive nuclear assistance. Superpowers, states with global power-projection capabilities, are threatened by nuclear proliferation anywhere in the international system, and they attempt to establish a hegemonic nonproliferation order to prevent sensitive nuclear assistance. States that do not depend on a superpower to provide for their security are less likely to fold under superpower pressure and are more likely to provide sensitive nuclear assistance.

The strategic theory of nuclear proliferation is supported by a battery of empirical tests. In chapter 2, a quantitative empirical analysis performed on an original sensitive nuclear assistance dataset strongly supported the strategic theory of nuclear proliferation. The suppliers' ability to project power over the recipient, the presence of a common enemy, and superpower dependence were statistically and substantively significant predictors of sensitive nuclear assistance.

In chapter 3, an in-depth case analysis provided further support for the strategic theory of nuclear proliferation. An examination of the power-projection capabilities of France and the United States, and of the corresponding strategic assessments made by French and U.S. officials, helped to account for the different approaches that these countries took to Israel's nuclear program from 1959 to 1965. France's inability to project conventional military power in the Middle East, an opportunity to constrain a key rival, and independence from the superpowers provided France with strategic incentives to assist Israel's nuclear weapons program. The U.S. ability to project power in the Middle East, however, led U.S. officials to a very different strategic assessment; U.S. officials feared the strategic consequences of nuclear proliferation in the Middle East because they had the ability to project military power in the region. As a result, the United States refused to help Israel's program and intervened, unsuccessfully, to prevent French-Israeli cooperation.

The chapter continued to present short studies of cases in which the potential nuclear suppliers and recipients were dependent on superpower pressure. These cases illustrate the importance of superpower dependence in shaping the probability of sensitive nuclear transfer. Analyses of thwarted nuclear cooperation between Argentina and Libya in 1985 and France and Taiwan in 1975 demonstrated that superpower-dependent states refrain from participation in sensitive nuclear transactions when they are confronted with superpower pressure.

Further qualitative analysis performed in chapter 4 explored the theory's limits. This chapter revealed that while the strategic theory of nuclear proliferation does not perfectly explain every case, it provides a firmer basis for understanding than rival explanations. An analysis of Chinese assistance to Pakistan in the 1980s demonstrated that the "enemy of my enemy is my customer" logic travels beyond French-Israeli nuclear cooperation. China provided sensitive nuclear assistance to Pakistan in order to impose strategic costs on its rival India.

Further, in this chapter we saw that the strategic theory of nuclear proliferation can even help to explain outlier cases. Despite its superpower status and the ability to project power over China, the Soviet Union provided sensitive nuclear assistance to China from 1958 to 1960. While the outcome of this case may be inconsistent with the expectations of the strategic theory of nuclear proliferation, the within-case analysis provides support for the theory. The Soviet Union was motivated to help China's nuclear program for a short period of time by a desire to constrain a common enemy: the United States. Furthermore, apart from the short-lived nuclear transfers and consistent with what we would expect from a superpower, the Soviet Union was generally opposed to nuclear proliferation in China.

This chapter also analyzed states that, according to the dictates of the theory, should have provided sensitive nuclear assistance but did not. An analysis of India and Israel demonstrated that the strategic conditions that encourage sensitive nuclear transfer were felt in policymaking circles in these states. Israel nearly provided sensitive nuclear assistance to South Africa in the 1970s, and strategic thinkers in New Delhi are arguing today that India should provide sensitive nuclear assistance to Taiwan and Vietnam in order to constrain Chinese power. We also saw, however, that as India's power-projection capability grows, New Delhi may become a less likely candidate to provide sensitive nuclear assistance over time. Next, a study of Pakistan's nuclear transfers to Iran, Libya, and North Korea demonstrated that the factors identified by the strategic theory of nuclear proliferation are powerful causes of sensitive nuclear assistance in the post–Cold War world. Finally, the chapter concluded with a consideration of how changes in power-projection capabilities in China and Russia are changing those states' propensities to provide sensitive nuclear assistance.

To answer the "So what?" question, chapter 5 clearly demonstrated the real-world importance of sensitive nuclear assistance. A large-N quantitative analysis, supplemented by qualitative case studies, showed that states that receive sensitive nuclear assistance are more likely to acquire nuclear weapons than states that do not. In short, chapter 5 demonstrated that sensitive nuclear assistance causes nuclear proliferation. This insight underscores the importance of understanding the research subject of this book: the causes of sensitive nuclear assistance.

Alternative explanations cannot account for the patterns of sensitive nuclear assistance. Economic explanations fall short. The quantitative analysis revealed that none of the economic variables are statistically and substantively significant predictors of sensitive nuclear assistance. The case studies also demonstrated the inadequacies of an economic approach. Economic considerations are often far from the minds of foreign policy officials involved in the decisions to export sensitive nuclear materials and technology. Indeed, the financial payoff from sensitive nuclear transfers is often insignificant. For this and other reasons, we saw that in cases of sensitive nuclear assistance, officials in the supplier states continually deliberate the strategic consequences of sensitive nuclear assistance, but the subject of economic incentives is rarely broached. This is not to say that economic considerations are irrelevant to state decisions to export sensitive nuclear materials and technology but that security considerations trump economic ones. States are perfectly willing to accept hard currency in exchange for sensitive nuclear transfers as long as the nuclear transfers are also consistent with an underlying strategic logic.

Neither do institutional or related normative arguments provide a satisfactory explanation for the patterns of nuclear transfer. The NPT and international norms against nuclear proliferation garner much attention, from policy analysts and academics alike. Indeed, while conducting my research, many analysts and scholars lectured me about how sensitive nuclear assistance was widespread in the early days of the nuclear era. They claimed that the ease with which sensitive nuclear assistance could be converted into a military nuclear program was not fully understood, that there were no international agreements prohibiting sensitive nuclear transfers, and that, unlike today, there was not a strong global norm against nuclear proliferation. Some even went so far as to argue that the research design of this book was fundamentally flawed because comparing nuclear proliferation today with nuclear proliferation in the past is like comparing apples and orangutans. While many scholars and analysts adhere to this position with deep conviction, there is little empirical support for it. In the case studies, we saw that in the pre-NPT era, there were states that freely provided sensitive nuclear assistance, while other states fiercely opposed it. Similarly, the case studies showed that in the post-NPT era, there were states that eagerly provided sensitive nuclear assistance, while other states vigorously resisted it. In addition, the statistical analysis revealed that sensitive nuclear assistance was not less likely in the post-NPT era. These results cast serious doubt on the argument that the establishment of the NPT led to a powerful norm against nuclear proliferation that restrains states from providing sensitive nuclear assistance.

The lack of a norm against sensitive nuclear assistance may be due in part to the conflicting messages in the NPT about the normative status of nuclear transfers. According to Article I of the treaty, nuclear weapon states agree "not in any way to assist, encourage, or induce any non-nuclear weapon state to manufacture or otherwise acquire nuclear weapons." On the other hand, Article IV of the treaty makes clear that "Nothing in this Treaty shall be interpreted as affecting the inalienable right of all the Parties to the Treaty to develop research, production and use of nuclear energy for peaceful purposes." Article IV continues, "All the Parties to the Treaty undertake to facilitate, and have the right to participate in, the fullest possible exchange of equipment, materials and scientific and technological information for the peaceful uses of nuclear energy."[1] But what does it mean for nuclear technology to be "peaceful"? Many sensitive nuclear technologies, such as uranium enrichment and plutonium reprocessing capabilities, have two uses. They can be used to produce fuel for a nuclear power reactor or for a nuclear weapon. In short, the dual-use nature of these key technologies combined with the norms articulated in Article IV of the NPT may even provide normative cover for those states that wish to promote the international spread of nuclear weapons.

In other words, the NPT may have failed to prevent sensitive nuclear transfers because it was not designed, strictly speaking, to prevent sensitive nuclear transfers. The NPT, however, may be better at doing what it was intended to do: slowing nuclear proliferation. In chapter 5, I found that NPT-member states have a lower risk of acquiring nuclear weapons than do non-NPT members. This relationship was evident even after controlling for other factors that influence nuclear proliferation. This suggests that the NPT does more than merely reflect the preferences of its member states and serves as an additional constraint on the international spread of nuclear weapons. This finding should provide some reassurance to advocates of the nuclear nonproliferation regime.

---

1. Treaty on the Nonproliferation of Nuclear Weapons, 1968.

Arguments that claim that states that possess nuclear weapons will be less likely to provide sensitive nuclear assistance do not find empirical support. In fact, if anything, the evidence appears to suggest that nuclear weapon states may be more likely to export sensitive nuclear materials and technology. If nuclear weapon states want to maintain an exclusive nuclear club, it is odd that they are so eager to coach prospective members through the admissions process.

The Causes of Nuclear Proliferation: A Supply-Side Approach

This book presents a new supply-side approach to understanding the causes of nuclear proliferation. The vast majority of the scholarly literature devoted to nuclear proliferation focuses on state demand for nuclear weapons. That is, existing scholarship painstakingly analyzes why states pursue or abandon nuclear programs. But this literature risks missing the point. At the end of the day, what we care about is not whether a country wants nuclear weapons, but whether it acquires them. In order for a state to acquire nuclear weapons, it is true that it must want them, but it must also be able to build them. And whether or not a state succeeds in constructing nuclear weapons is heavily dependent on the actions that other states take to assist or impede them. With the exception of a few advanced industrialized countries, a state's ability to build nuclear weapons generally hinges on its ability to find an international supplier. Many states were only able to develop nuclear arsenals and advanced nuclear weapons programs because they received some form of external assistance from more advanced nuclear states. Contrariwise, some states were prevented from acquiring nuclear weapons because other states applied pressure on them in the form of technology denial, sanctions, or preventive military strikes. If we are to understand how and why nuclear weapons spread, it is necessary to understand the supply side of nuclear proliferation.

Unfortunately, we lack at present a theoretical answer to the most basic supply-side questions: why do some states support nuclear proliferation to additional states, while other states oppose it? Why are some states willing to apply tough sanctions on states that are developing nuclear weapons programs? Why do some states support preventive military strikes against other states' nuclear facilities? What determines variation in the way states vote on nuclear proliferation issues in the IAEA Board of Governors or in the UN Security Council? Why do some states provide sensitive nuclear assistance to nonnuclear weapon states? These are all interesting questions with theoretical and empirical significance. They are also questions that have not yet been systematically addressed by international relations scholars, but they could become the focus of future scholarly analysis. For example, scholars could examine the conditions under which states launch preventive military strikes against other states' nuclear facilities. This question is becoming increasingly important in the wake of Israel's strike on a nuclear reactor in Syria in 2007 and amid widespread speculation that the United States might target Iranian nuclear facilities in a future military operation.

In fact, the strategic theory of nuclear proliferation presented here may provide a set of hypotheses to answer this diverse set of questions. For example, Hypothesis 1 would predict that a state's ability to project power over a particular state may determine whether, and the degree to which, it will oppose nuclear proliferation to that state. States that enjoy the ability to project conventional military power over a particular state should be more threatened by nuclear proliferation to that state. Thus, power-projecting states may be more likely to vote against the state in the IAEA Board of Governors and the UN Security Council, more likely to impose sanctions in an effort to dissuade the state from its nuclear course, more likely to support preventive military strikes against the state's nuclear facilities, and less likely to provide the state with sensitive nuclear assistance. On the other hand, states that lack the ability to project power over a particular state will be less threatened by proliferation to that state. Therefore, we may expect non-power-projecting states to be less likely to vote against the state on measures related to the state's nuclear program in international institutions, less likely to impose tough sanctions designed to strangle a state's nuclear program, less likely to support preventive military strikes against the state's nuclear facilities, and more likely to provide sensitive nuclear assistance to the state. Of course, there are undoubtedly many other factors that influence the degree to which states are threatened by nuclear proliferation and how far they are willing to go to stop it. This book provides the first theoretical explanation and empirical test of the causes of sensitive nuclear assistance. Future research should examine the applicability of these and alternative hypotheses to other important issues related to the supply of nuclear proliferation.

The Differential Effects of Nuclear Proliferation

This book speaks to the high-profile debate on the consequences of nuclear proliferation. The debate, as it stands today, pits those who argue that "more may be better" against those who argue that "more is worse."[2] This debate has long felt unsatisfying to many scholars and practitioners. The Waltzian argument that

---

2. Sagan and Waltz, *Spread of Nuclear Weapons*.

increasing numbers of countries in possession of nuclear weapons may actually be a good thing is, to many analysts, simply implausible. Clearly, the spread of nuclear weapons (any one of which could destroy an entire city) cannot, on balance, be a positive development. On the other hand, Sagan's argument about nuclear proliferation increasing international instability through an increased risk of nuclear accident, preventive strikes, and crisis instability is more plausible—but incomplete. Can the widespread nuclear nonproliferation sentiment found in many national governments be explained solely by the fact that governments worry about international instability? The answer is probably not. Government officials do not make policy with the primary aim of contributing to the stability of the international system; rather, officials pursue policies that serve the interests of their own state.[3]

The argument presented in this book occupies a much-needed middle ground in this debate. Waltz argues that the spread of nuclear weapons is good, Sagan argues that it is bad, and I argue that it depends: *the spread of nuclear weapons is bad for power-projecting states and can be good for non-power-projecting states.* As nuclear weapons spread, the military freedom of action of power-projecting states is constrained. This is bad for power-projecting states because they are less able to use their conventional military power to secure their interests. On the other hand, nuclear proliferation can improve the strategic environment of non-power-projecting states, even if they lack nuclear weapons themselves. As the spread of nuclear weapons constrains power-projecting states, non-power-projecting states may become less vulnerable to powerful states using conventional forces in a way that potentially threatens their interests.

A simple thought experiment provides a useful illustration. Imagine a world completely devoid of nuclear weapons. Just think of the international system prior to 1945. In this world, conventional military power is a very valuable currency. States can use military might to protect their interests, and weak states are often at the mercy of their more powerful brethren. Now, imagine a very different world, one in which every single state in the international system possesses nuclear weapons. The effective use of conventional military power becomes much more difficult and, arguably, irrelevant in this world. States are often deterred from using large-scale military force for fear of nuclear retaliation. The currency of conventional military power is devalued. Powerful states cannot easily use military might to secure their interests. Weak states are liberated from having their interests continually threatened by the use, or the threatened use, of

---

3. Of course, states may sometimes have an interest in promoting international stability, but policymakers do not generally view international stability as a good, in and of itself, that trumps the national interest.

conventional military power by more powerful states. The advantage that conventional military powers have traditionally enjoyed largely evaporates in this world. A completely proliferated world may present a dream world for some weak states, but it is a nightmare scenario for powerful states.[4]

Some may question why, if nuclear proliferation is less threatening to non-power-projecting states, some small powers vigorously oppose the proliferation of nuclear weapons. After all, Ireland was the first country to ratify the NPT, New Zealand is extremely proud of its anti-nuclear stance, and many other small powers actively support international nuclear nonproliferation efforts. The answer is that understanding a state's power-projection capability cannot tell you everything about a state's proliferation preferences, but it is a useful place to start. For power-projecting states, a fear of nuclear proliferation is dictated by their position in the international system. The structural condition of non-power-projecting states, on the other hand, allows them greater flexibility. Some non-power-projecting states, like New Zealand, may vigorously oppose nuclear proliferation, while others like Pakistan will help other states acquire nuclear weapons. In order to understand whether a non-power-projecting state will benefit or be threatened by the spread of nuclear weapons, other factors must be taken into consideration. Some of these variables, such as a shared enemy with a potential nuclear weapon state and dependence on a superpower patron, have been identified in this book. Uncovering other relevant factors should become the goal of future research.

Indeed, I hope that the novel approach presented in this book may reinvigorate the scholarly study of the consequences of nuclear proliferation by establishing a research agenda on the differential effects of nuclear proliferation. Future studies could identify other unit-level effects of nuclear proliferation and examine the factors, other than power projection, that shape the degree to which states will be threatened as nuclear weapons spread.

Balance of Threat Theory

The analysis presented in this book may also suggest a useful extension of balance of threat theory. Balance of threat theory argues that states band together in international alliances to counter shared threats. As demonstrated by the title of the landmark book on the subject, *The Origins of Alliances* by Stephen Walt,

---

4. For another thought experiment on politics in a proliferated world, see Morton Kaplan's writing on "unit-veto systems" in *System and Process in International Politics* (London: John Wiley and Sons, 1965).

balance of threat theory seeks to explain the patterns of international alliances.[5] When I began this research, I speculated that states may be more likely to provide sensitive nuclear assistance to allied states. In other words, I assumed that sensitive nuclear assistance may be one of a larger set of behaviors occurring within an alliance framework and that sensitive nuclear assistance may simply be a type of balancing behavior. To my surprise, upon closer examination, I discovered that none of the cases of sensitive nuclear assistance occurred between states that shared a formal alliance. Of course, many of the states in dyads in which nuclear assistance took place enjoyed friendly relations, and in no case were they mortal enemies. Nevertheless, it is shocking that the formation of alliances and the provision of sensitive nuclear assistance appear to be mutually exclusive behaviors. Many of the cases of nuclear cooperation occurred between states that faced shared threats, as demonstrated by the empirical support found for Hypothesis 2. Yet, contrary to the expectations of balance of threat theory, these states did not choose to band together in international alliances. Instead, they chose to counter the shared threat by transferring sensitive nuclear materials and technology. This finding may suggest that transfers of sensitive nuclear materials and technology (and perhaps arms sales more broadly) are a method of external balancing logically and empirically distinct from alliance formation. In certain cases, states appear to prefer to counter a shared threat through means other than the formation of alliances. This may be because alliance obligations often carry significant downsides, including the possibility of being drawn into an unwanted war.[6] Sensitive nuclear assistance, on the other hand, can be an equally, if not more, effective method for countering a shared threat, while simultaneously allowing states to avoid the entanglements of formal military alliances. Further scholarly research could explore the conditions under which states prefer to balance threats with alliances as opposed to using other options, including sensitive nuclear assistance.

Conventional Arms Sales

The theoretical approach of this book may provide a helpful framework for the study of conventional arms transfers.[7] Much of the research on conventional

---

5. Stephen Walt, *The Origins of Alliances* (Ithaca: Cornell University Press, 1990).
6. See, for instance, Snyder, "The Security Dilemma in Alliance Politics," pp. 461–495.
7. On conventional arms sales, see for example, Pierre, *Global Politics of Arms Sales*; and Shannon Lindsey Blanton, "Promoting Human Rights and Democracy in the Developing World," *American Journal of Political Science*, Vol. 44, No. 1 (January 2000), pp. 123–131.

arms sales has focused on the effects of arms transfers on regional stability.[8] The smaller body of research that has examined the causes of arms transfers has tended to focus on examining the behavior of specific arms suppliers. For example, Shannon Lindsay Blanton has examined the relationship between human rights and democracy in U.S. arms export policy and has found that the United States is more likely to transfer arms to democratic countries.[9] Blanton's research has made an important contribution to our understanding of the correlates of U.S. weapons exports, but it cannot tell us whether these findings are country-specific, or whether they are applicable to arms suppliers more broadly. Perhaps the United States is alone in favoring democratic states, and other arms suppliers are less concerned about the regime type of the recipient state. Indeed, this book finds that power-based factors are paramount and that the regime type of the recipient does not affect decisions to transfer sensitive nuclear materials and technology.[10] Scholars could seek to understand further the conditions under which states export conventional weapons by systematically examining the entire universe of potential suppliers and by including the strategic factors identified here among the explanatory variables. It is possible that power-projection capabilities, the presence of a common enemy, and superpower dependence shape the probability that states will transfer conventional military, as well as nuclear, materials and technology.

Furthermore, nuclear and conventional military technologies are thought to have different effects on the nature of warfare that could affect their supply patterns. According to nuclear deterrence theory, when a state acquires nuclear weapons, any ability that other states had to use large-scale military force against it is largely eliminated.[11] For example, if Egypt acquired nuclear weapons, the United States would likely be deterred from using military force against Egypt for fear of nuclear retaliation. When a state acquires a new conventional military capability, however, other states will find it more difficult to use military power against it, but it is unlikely that they would be completely deterred from doing so. If Egypt acquired a new squadron of fighter aircraft, it may be more difficult for the United States to achieve air superiority in a military conflict with Egypt,

---

8. On the relationship between arms transfers and conflict, see, for instance, David Kinsella and Herbert K. Tillema, "Arms and Aggression in the Middle East," *Journal of Conflict Resolution*, Vol. 39, No. 2 (June 1995), pp. 306–329; and Gregory Sanjin, "Promoting Stability or Instability?" *International Studies Quarterly*, Vol. 43, No. 4 (December 1999), pp. 641–670.

9. Blanton, "Promoting Human Rights"; and Shannon Lindsey Blanton, "Foreign Policy in Transition?" *International Studies Quarterly*, Vol. 49, No. 4 (December 2005), pp. 647–667.

10. See chapter 2 for the findings on the lack of a relationship between the regime type of the recipient and the probability of sensitive nuclear assistance.

11. On nuclear deterrence theory, see, for example, Schelling, *Arms and Influence*; Jervis, *Meaning of the Nuclear Revolution*; and Powell, *Nuclear Deterrence Theory*.

but a few new jets alone would be unlikely to deter the United States from using military force. In other words, incremental increases in the conventional military power of one state may make it slightly more difficult for other states to use military force against it, but nuclear weapons may completely deter other states from even considering the military option.

This fundamental difference in the impact of conventional and nuclear weapons on military conflict may have two implications for the motivations of the potential suppliers of these technologies. First, in contrast to the findings of the literature on conventional arms sales, sensitive nuclear assistance is not driven by a desire to gain greater influence over the foreign policy of the recipient state. Some have argued that states export conventional military hardware to gain a voice in the formation of the recipient's foreign policy.[12] If the recipient becomes dependent on a particular supplier state for the supply of its weapons, it may be more open to considering that state's demands. Nuclear and conventional arms, however, may be quite different in this regard. Scholars have argued that nuclear weapons provide states with security independence.[13] If this is correct, nuclear assistance resulting in nuclear proliferation should actually serve to reduce, not increase, the supplier state's influence over the recipient state. Conventional arms sales may be driven, in part, by a desire to gain influence over the recipient, but this book has demonstrated that the provision of sensitive nuclear materials and technology is driven by a different logic.

Second, relative power between supplier and recipient may be an important determinant of arms transfers, but the direction of its effect may change from the nuclear to the conventional realm. When it comes to conventional arms sales (and in sharp contrast with sensitive nuclear transfers), it may be the most powerful states that have the greatest strategic incentives to supply. A state can provide conventional military technologies to another state without seriously threatening its own strategic positions as long as it is able to retain a sizable conventional military superiority over the recipient state. States that are most powerful in relation to a particular recipient have a larger margin of superiority within which to operate and may have more leeway to export conventional arms without incurring strategic costs. On the other hand, less powerful states may be more threatened by the spread of conventional military technology because they are more vulnerable to shifts in the conventional military balance between themselves and the recipient. For this reason, we may expect that powerful states may be more likely to sell conventional arms than weak states. A brief look at the

---

12. For the argument that states trade conventional arms for influence, see Pierre, *Global Politics of Arms Sales;* and John Sislin, "Arms as Influence," *Journal of Conflict Resolution,* Vol. 38, No. 4 (December 1994), pp. 655–689.

13. Weber, "Cooperation and Interdependence."

empirical evidence appears to validate this expectation. In a reversal of the patterns of sensitive nuclear assistance, it is superpowers that are the most prolific exporters of conventional arms.[14] Of course, there are other reasons why powerful states may be more likely to trade in conventional arms—that they may have a greater supply of military hardware to export, for starters. Further research could refine the logic of this hypothesis and test it systematically against possible alternatives.

Understanding Transnational Security Threats

The basic logic underlying the strategic theory of nuclear proliferation may shed light on state responses to transnational security threats other than nuclear proliferation, including terrorism, climate change, and global pandemics. These problems may be becoming increasingly important, as scholars have argued that the number and severity of transnational security threats may be increasing and that this may be due, in part, to the current unipolar distribution of power in the international system.[15]

It is often claimed that since these new transnational security issues threaten the entire international system, they offer the promise of widespread international cooperation. For example, Allen Weiner has argued that "there is an underlying affinity of interests among the Permanent Members [of the UN Security Council] with respect to these [contemporary security] threats."[16] Similarly, in a call for "consensual security," John Steinbruner claims that nuclear proliferation, global climate change, and terrorism, among other challenges, pose a threat to the global order and concludes: "Security policies responsive to the emerging circumstances of globalization would make the defense of global legal order its primary objective and would therefore elevate common interest over national advantage."[17] This idea that great powers will work together to fight common security threats is also consistent with the expectations of much of realist international relations theory. While realists generally believe that sustained great power cooperation will be unlikely in an anarchic system, they also argue that nothing is better than a shared security threat to bring great powers together.

---

14. David Kinsella, "Conflict in Context," *American Journal of Political Science*, Vol. 38, No. 3 (August 1994), pp. 557–581.
15. Weber, Barma, Kroenig, and Ratner, "How Globalization Went Bad."
16. Allen Weiner, "The Use of Force and Contemporary Security Threats" *Stanford Law Review*, Vol. 59, No. 2 (2006), pp. 415–504.
17. John Steinbruner, "Consensual Security," *Bulletin of Atomic Scientists*, Vol. 64, No. 1 (March/April 2008), pp. 23–27.

The argument of this book, however, gives us reason to be skeptical about the possibility of great power cooperation to confront global challenges. The hope of international cooperation rests on a superficial assessment of how common these threats truly are. Terrorism, pandemics, global climate change, and nuclear proliferation may very well threaten all states to some degree, but they almost certainly threaten some states more than others. This book has shown that states vary in the degree to which they are threatened by nuclear proliferation, resulting in uneven interest and attention to the nuclear proliferation threat. Similarly, states that are less threatened by other transnational security threats, such as terrorism, climate change, or global pandemics, will be less motivated to cooperate to find a solution. For example, a recent U.S. National Intelligence Council report suggests that in the coming years China will be less threatened by radical Islamic extremism than the United States, Europe, or Russia and that, for this reason, it will be difficult for the United States to get sustained Chinese assistance on issues of counterterrorism.[18] Scholars and policy analysts must examine both the systemic and the differential costs of transnational security threats if they hope to understand the conditions under which great powers will cooperate to address them.

Implications for U.S. Foreign Policy

The argument and findings of this book contain lessons for U.S. nuclear nonproliferation policymakers. The U.S. foreign policy community is currently seeking to develop a policy that would deter the transfer of sensitive nuclear materials and technology. As was discussed in the introduction, following North Korea's nuclear test in October 2006, President Bush threatened that the United States would hold North Korea responsible if it were to export sensitive nuclear materials and technologies to states or terrorist networks.[19] Many analysts have argued that the United States should generalize this deterrent threat and promise to punish any country that provides sensitive nuclear materials and technology.[20] Others have begun developing a nuclear forensics capability that would provide the United States and the international community with the ability to attribute the likely source of nuclear material following a nuclear attack.[21] A robust nuclear forensics

---

18. National Intelligence Council, 2020 Project, "Mapping the Global Future," December 2004.
19. "President Bush's Statement on North Korea Nuclear Test."
20. See, for instance, Robert L. Gallucci, "Averting Nuclear Catastrophe," *Harvard International Review*, Vol. 26, No. 4 (Winter 2005); and Michael Levi, *On Nuclear Terrorism* (Cambridge: Harvard University Press 2007).
21. See, for instance, William Dunlop and Harold Smith, "Who Did It?" *Arms Control Today*, Vol. 36, No. 8 (October 2006); Caitlin Talmadge, "Deterring a Nuclear 9/11," *Washington Quarterly*,

capability could help the United States to determine the origins of the material used in a nuclear explosion, enhancing the credibility of the U.S. threats to retaliate against the nuclear supplier. At the same time, the 2006 National Security Strategy of the United States of America disavows a one-size-fits-all deterrence policy and pledges to develop a policy of "tailored deterrence" crafted to fit specific adversaries and specific threats.[22] In order to design a tailored policy to deter specific states from providing sensitive nuclear assistance, foreign policymakers and intelligence analysts must first identify the states that are at the greatest risk of providing sensitive nuclear assistance. So the first question that a good intelligence analyst should ask is: who is next? Which countries are most likely to provide sensitive nuclear assistance?

This book provides the answer to this question. Analysts often look to states that would have economic motivations for selling sensitive nuclear materials and technology, fingering North Korea, Russia, and Pakistan as the most likely nuclear suppliers. We have seen, however, that economic conditions bear very little on state decisions to provide sensitive nuclear assistance. Instead, analysts must devote their attention to the strategic conditions that contribute to sensitive nuclear exports. The states that are most likely to export sensitive nuclear materials and technology are states with relatively circumscribed force-projection capabilities, states that do not rely on a U.S. security guarantee, and states that could gain an advantage by imposing strategic costs on a powerful enemy. When one considers these strategic factors, the list of at-risk states includes some states that are already seen as likely suspects, like North Korea, Russia, and Pakistan. But it also includes states that are currently considered reliable U.S. allies, such as India, France, and Israel. As we saw in chapter 4, India may have strategic incentives to provide sensitive nuclear assistance to Vietnam and Taiwan, and some Indian strategists are currently arguing that India should exercise this option.[23] France has repeatedly provided sensitive nuclear assistance to nonnuclear weapon states over U.S. opposition. And Israel provided strategic military aid to South Africa at a time when the United States was working to build international pressure to isolate and weaken Pretoria. U.S. officials must devote appropriate attention to the full list of at-risk states and accurately assess the strategic factors that are the true drivers of sensitive nuclear transfers.

This advice extends beyond the realm of nuclear transfers to questions of nuclear nonproliferation more broadly. The United States often struggles to

---

Vol. 30, No. 2 (Spring 2007), pp. 21–34; and Daniel H. Chivers, Bethany F. Lyles, Brett H. Isselhardt, and Jonathan S. Snyder, "Before the Day After," *Arms Control Today* (July/August 2008).

22. "National Security Strategy of the United States of America," March 2006, p. 43.

23. However, the probability that India will export sensitive nuclear materials and technology may decline if India's conventional military power continues to grow.

convince other great powers to join it in its efforts to pressure countries that are pursuing nuclear weapons capabilities. Policymakers in Washington are often puzzled that it is so difficult to get international cooperation on nuclear nonproliferation issues. Living in a world (the Washington, D.C., beltway) in which nuclear proliferation is demonized, they cannot imagine how officials in other capitals, such as Beijing and Moscow, are not horrified by the thought of nuclear weapons in Tehran or Pyongyang. When states are unwilling to press another state on its nuclear program, officials in Washington often assume that foreign officials do not properly understand the threat posed by nuclear proliferation. U.S. officials believe the excuses that foreign governments give them for their reluctance to press harder: that the nuclear proliferator in question is technically incapable of producing nuclear weapons, that time is on our side because it will be many years before the potential proliferator's nuclear development reaches the point of no return, or that the state in question is not genuinely interested in nuclear weapons and is merely pursuing a civilian capability. Then the Kabuki dance begins. U.S. officials embark on a campaign to convince other governments of the threat. They provide intelligence in an attempt to demonstrate that the proliferator in question is making serious progress on its nuclear program. They present more intelligence to demonstrate that the proliferator in question likely has a military option in mind. Then they point out that the likely proliferator is actually geographically more proximate to the foreign officials' capitals than it is to Washington, so that they should actually feel more threatened than the United States by the prospect of nuclear weapons in that state. When the United States has rested its case and the foreign governments are still unwilling to take tough measures to stop proliferation, Washington chalks it up to economic incentives. U.S. officials assume that foreign governments are unwilling to push the potential nuclear proliferator on its nuclear program because they do not want to jeopardize existing economic relationships with that country.

U.S. officials need to see the difficulty of getting international nuclear nonproliferation cooperation for what it is: *nuclear proliferation threatens the United States more than any other state on the globe.* The United States is a global superpower, and nuclear proliferation anywhere is a threat to America's strategic position. For other states with more limited spheres of influence, nuclear proliferation in a distant region is not a threat. In fact, these countries may even see a significant upside to the spread of nuclear weapons—because nuclear proliferation means a constrained and thus a weakened United States. The reluctance of foreign governments to bear a burden to stop proliferation in a distant region does not happen because they misunderstand the threat; it happens because they understand it perfectly well. For them, it is not a problem. The failure of understanding is on the U.S. side. The United States will continue to fail to convince other powers to join in a fight against nuclear proliferation that disproportionately threatens the United States—until it realizes the true dynamics at work.

Some may question whether the argument of this book is damaging to international nuclear nonproliferation efforts. By pointing out that nuclear proliferation disproportionately threatens powerful states, the argument and evidence presented in this book could encourage leaders in relatively weak states to be less concerned about their commitments to stop nuclear proliferation globally and even convince them to provide sensitive nuclear assistance to other states. To those who would draw such a conclusion from this book, a warning is in order. Sensitive nuclear assistance is a permanent solution to a temporary problem. As we have seen, states often provide sensitive nuclear assistance in order to constrain a key rival, but patterns of friendship and enmity are fluid in international politics. The policymakers who authorized sensitive nuclear transfers rarely considered the long-term consequences of their behavior at the time of the decision. Leaders often acted on a short time horizon because they were dealing with a pressing security crisis, and later came to regret their decisions. For example, just a few years after the Soviet Union provided sensitive nuclear assistance to Beijing, Moscow became so threatened by the specter of nuclear proliferation in China that it considered a preventive military strike to destroy China's nuclear facilities. Leaders in nuclear-capable states should avoid repeating this mistake. While helping other states acquire nuclear weapons may appear to make strategic sense in the short run, policymakers should seriously consider the long-run consequences before contributing to the spread of the world's most dangerous weapons.

Finally, the topic that has been on everyone's mind since September 11, 2001: nuclear terrorism.[24] Osama bin Laden has stated that it is a religious duty for Islamic terrorists to acquire nuclear weapons, and there is evidence that Al-Qaeda has actively sought to develop a nuclear capability.[25] Radical Muslim clerics have issued *fatwas* declaring that the use of nuclear weapons against infidels is consistent with the teachings of the Koran.[26] If Al-Qaeda ever acquired nuclear weapons, most terrorism analysts believe that they would use them.[27] Accordingly,

---

24. On nuclear terrorism, see Graham Allison, *Nuclear Terrorism* (New York: Times Books, 2005).
25. Bob Woodward, Robert G. Kaiser, David B. Ottaway, "U.S. Fears Bin Laden Made Nuclear Strides," *Washington Post*, December 4, 2001, A1.
26. Colin Freeman and Philip Sherwell, "Iranian Fatwa Approves Use of Nuclear Weapons," *The Telegraph* (London), February 19, 2006.
27. For a review of the debate and a view that Al-Qaeda might opt not to use nuclear weapons if it acquired them, see Lewis Dunn, "Can Al Qaeda Be Deterred from Using Nuclear Weapons?" *Center for the Study of Weapons of Mass Destruction Occasional Paper, No. 3* (Washington, D.C.: National Defense University Press, July 2005).

many analysts believe that nuclear terrorism poses the greatest threat to U.S. national security.

Of course, it would be nearly impossible for a terrorist organization to construct nuclear weapons from scratch. Terrorist organizations lack the infrastructure and expertise required to construct an indigenous nuclear weapons production program. To acquire nuclear weapons, therefore, most experts believe that a terrorist network would either have to steal sensitive nuclear materials and technology or have states provide them with sensitive nuclear assistance.[28]

Would states provide sensitive nuclear assistance to terrorist organizations? The findings of this book suggest that the prospect of sensitive nuclear cooperation between states and terrorist organizations is highly unlikely. Nuclear weapons and terrorist organizations have existed side by side for the past sixty years. During this time, states have repeatedly supplied sensitive nuclear assistance to other states but have never given sensitive nuclear assistance to terrorists. Why? The findings of this book may provide some partial answers. We have seen that states maintain careful control over their most sensitive nuclear technologies and materials and that the substate smuggling of sensitive nuclear materials and technology without the state's approval or knowledge has historically been quite rare. So, while many terrorist watchers worry about rogue scientists or other substate actors helping terrorists to acquire nuclear weapons, it is unlikely that rogue actors would be able to provide terrorist organizations with levels of support sufficient for making nuclear weapons without the government's knowledge and authorization. Furthermore, we have seen that states are very careful to limit the provision of sensitive nuclear assistance to situations that do not directly threaten their own strategic position. When providing sensitive nuclear materials or technologies to terrorist organizations, however, there are fewer guarantees that the nuclear assistance will not come back to haunt the supplier. Terrorists could decide to use the nuclear weapons to blackmail the supplier. They could turn on the supplier state and attack it with its own weapons. Terrorists could strike a third state with nuclear weapons and this third state could then decide to retaliate against the supplier state. For these reasons (and probably many others) helping terrorists to acquire nuclear weapons is a risky proposition. It is unlikely that states will provide sensitive nuclear assistance to a terrorist organization in the foreseeable future.

The strategic incentives that drive states to provide sensitive nuclear assistance to other states, however, will continue to contribute to the international spread of nuclear weapons.

---

28. See for instance, Matthew Bunn and Anthony Wier, "Terrorist Nuclear Weapon Construction" *Annals of the American Academy of Political and Social Science* (2006).

APPENDIXES

Appendix A: Data Appendix for Chapter 2
Independent Variables

To assess the effects of economic motivations on state decisions to provide sensitive nuclear assistance, I include a number of economic control variables. *Economic development* is measured as a country's GDP per capita in constant 1996 dollars. Economic *Growth* is calculated as growth = $\log(\text{GDP}_t) - \log(\text{GDP}_{t-1})$. Following Oneal and Russett, I measure *Trade dependence* as total trade (imports plus exports) between the two member states of a dyad as a percentage of the GDP of the first state.[1] *Openness* to international trade is calculated as a state's trade ratio, total trade (imports plus exports) divided by GDP. The openness variable draws on data from Singh and Way.[2] All other economic data are from Kristian Gleditsch and extracted using EUGene.[3]

I also control for a number of institutional variables that could affect patterns of international nuclear trade. The institutions of the nuclear nonproliferation regime set restrictions on the transfer of nuclear materials and technology, which may render member states less likely to provide sensitive nuclear assistance. To measure the effect of international institutions on sensitive nuclear assistance,

---

1. John R. Oneal and Bruce Russett, "The Classical Liberals Were Right," *International Studies Quarterly*, Vol. 41, No. 2 (June 1997), pp. 267–294.
2. Singh and Way, "Correlates of Nuclear Proliferation."
3. Kristian Gleditsch, "Expanded Trade and GDP Data," *Journal of Conflict Resolution*, Vol. 46, No. 5 (2002), pp. 712–724; and Bennett and Stam, "EUGene."

I construct three dichotomous variables. *NPT* measures whether a supplier state is a member of the NPT.[4] *NPT2* indicates whether the potential recipient is a member of the NPT. *NSG* measures whether the potential supplier is a member of the NSG.[5] Previous research suggests that domestic regime type may affect a state's behavior on nuclear weapons issues.[6] To measure *Regime type,* I use polity scores which range from −10 (most autocratic) to +10 (most democratic) from the Polity IV dataset.[7]

We may expect states, regardless of their level of conventional military power, to be reluctant to provide sensitive nuclear assistance to geographically proximate states. To control for this factor, I generate *Distance,* a logged measure of the number of miles between capital cities, as calculated by EUGene.[8] It is likely, however, that the relationship between distance and sensitive nuclear assistance is non-monotonic. Previous analysis has suggested that logistical problems may make it difficult for states to transfer sensitive nuclear technologies to distant states.[9] To test for a non-monotonic relationship between distance and sensitive nuclear transfers, I also include *Distance squared,* a squared term of the distance variable.[10]

One may expect that states will be more likely to provide sensitive nuclear assistance to allied states. As I argued in chapter 1, however, an alliance variable is not included in the statistical models, because contrary to the expectation of this alliance-building hypothesis, there is no instance of a state providing sensitive nuclear assistance to a state with which it shared a formal alliance.

I also control for factors that influence the demand side of nuclear transactions. Previous research suggests that economic development, openness to the international economy, economic liberalization, and threat environment shape

---

4. The NPT, established in 1968, is the cornerstone of the nuclear nonproliferation regime. Information on membership in the NPT is from the Institute for Defense and Disarmament Studies, *Nonproliferation Treaty.*

5. The Nuclear Suppliers Group is a nuclear cartel founded in 1975 composed of states with advanced nuclear capabilities. Information on the Nuclear Suppliers Group information is taken from Tadeusz Strulak, "The Nuclear Suppliers Group," *Nonproliferation Review,* Vol. 1, No. 1 (Fall 1993), pp. 2–10; Nuclear Suppliers Group, "Guidelines for Nuclear Transfers"; and Nuclear Threat Initiative, *Country Profiles.*

6. Singh and Way, "Correlates of Nuclear Proliferation."

7. K. Jaggers and T. R. Gurr, "Tracking Democracy's Third Wave with the Polity III Data," *Journal of Peace Research,* Vol. 32, No. 4 (1995), pp. 469–82.

8. Bennett and Stam, "EUGene."

9. Amy Sands, "Emerging Nuclear Suppliers," in Potter, ed., *International Nuclear Trade and Nonproliferation.*

10. Including a variable and its squared term in a statistical model is a common method for testing for a non-monotonic relationship. See Ramsey Schafer, *The Statistical Sleuth* (Pacific Grove, CA: Duxbury, 2002), pp. 244–245.

a state's demand for nuclear weapons.[11] Thus, it is likely that these variables may also shape a recipient's demand for sensitive nuclear assistance. We may also expect that, like nuclear supplier states, potential nuclear recipient states that are dependent on a superpower patron may be more vulnerable to superpower pressure and will be less likely to receive sensitive nuclear assistance. The indicators of economic development, openness, and superpower dependence have already been discussed. To measure *Liberalization,* I use a variable from Singh and Way that gauges the movement toward greater trade openness by calculating the change in openness over time spans of three, five, and ten years.[12] The security environment of the recipient is measured using *Disputes.* It is a five-year moving average of the number of militarized interstate disputes per year in which a recipient state is involved. This measure is from Singh and Way and draws on data from version 3.0 of the Militarized Interstate Dispute data set.[13]

Further Robustness Checks

I supplemented the statistical analysis reported above with robustness checks to examine the extent to which my results depend on the coding of the universe of cases and the dependent variable, country-specific factors, model specification, or statistical technique.

To begin, I experimented with alternate codings of the universe of cases. It is difficult to define precisely when some states acquired the capability to become a nuclear supplier and thus when exactly a state should enter the analysis as a potential supplier state. It is also a challenge to specify exactly when some countries acquired nuclear weapons and thus drop out of the analysis as potential nuclear recipients. For example, it is widely believed that North Korea possessed enough fissile material to produce a nuclear weapon in the early 1990s, but experts disagree when, or even if, North Korea produced a functioning nuclear device.[14] Alternate codings of the North Korea case, and others like it, reveal that the results reported above are not sensitive to alternate codings of the universe of cases.

---

11. See, for example, Singh and Way, "Correlates of Nuclear Proliferation"; and Dong-Joon Jo and Erik Gartzke, "Determinants of Nuclear Weapons Proliferation," *Journal of Conflict Resolution,* Vol. 51, No. 1 (February 2007), pp. 167–194.
12. Singh and Way, "Correlates of Nuclear Proliferation."
13. Singh and Way, "Correlates of Nuclear Proliferation." On version 3.0 of the Militarized Interstate Dispute data set, see Faten Ghosen and Glenn Palmer, *Codebook for the Militarized Interstate Dispute Data,* Version 3.0. 2003.
14. North Korea tested nuclear devices in October 2006 and May of 2009, but many experts considered the tests failures and evidence that North Korea lacks the capability to produce a functioning nuclear warhead.

I also explored the sensitivity of the results to alternate codings of the dependent variable. In the first test, I expanded the definition of sensitive nuclear assistance to include instances of nuclear cooperation listed in appendices D and E that do not qualify as sensitive nuclear assistance, but that may nevertheless have contributed to nuclear proliferation. The additional cases of nuclear assistance include U.S. exports to India in 1955, Soviet aid to North Korea in 1965, and Russia's provision of nuclear facilities to Iran from 1995 to the present day. In a second test, I expanded the definition of sensitive nuclear assistance to include cases in which a capable nuclear supplier agreed to provide sensitive nuclear assistance, but, for whatever reason, did not execute the sensitive nuclear transaction. Cases of canceled sensitive nuclear transactions include Argentina's offer to transfer reprocessing technology to Libya in 1985, France's contract to provide uranium enrichment capabilities to Australia in 1969, France's agreement to export reprocessing facilities to South Korea in 1975 and 1976, Russia's offer to provide uranium enrichment technology to Iran in 1995, and Pakistan's offer of nuclear aid to Iraq and Syria in the mid-1990s. Expanding the coding of sensitive nuclear assistance to include this broader range of cases and reestimating the models did not alter the findings.

To examine whether the results are driven by the nuclear export behavior of specific nuclear suppliers, I dropped dyads containing certain key countries and repeated the analysis. Sequentially removing the dyads containing China, France, Pakistan, and the United States, and reestimating the models did not change the results.

Next, to ensure that my results were not being driven by the inclusion of specific control variables, I reran scores of models sequentially omitting right-hand-side variables one at a time. Again, the core results were not affected.

As a final test, I reestimated the models, using nonparametric matching techniques, as recommended by Ho et al.[15] I performed three separate matching analyses with each of the three key independent variables (*Power projection, Enemy, Superpower pact*) taking a turn as the treatment. In order to include *Power projection* as the treatment, I dichotomized the variable, recoding scores equal to or greater than zero as one and recoding scores less than zero as zero. To preprocess the data, one-to-one nearest neighbor matching with replacement was employed, using GenMatch.[16] I then repeated the parametric analysis, using

---

15. On nonparametric, matching techniques, see Daniel Ho, Kosuke Imai, Gary King, and Elizabeth A. Stuart, "Matching as Non-parametric Preprocessing for Reducing Model Dependence in Parametric Causal Inference," *Political Analysis,* Vol. 15 (2007), pp. 199–236.

16. Jasjeet S. Sekhon, "Multivariate and Propensity Score Matching Software with Automated Balance Optimization," *Journal of Statistical Software,* forthcoming; Jasjeet S. Sekhon, "Matching,"

ReLogit. The core findings were unaltered. Next, I applied a caliper that dropped observations that lacked sufficiently close matches, which I defined as observations that were more than one standard deviation away from their matched pair on any covariate. Again, I repeated the parametric analysis; the key results were not affected.

Appendix B: Data Appendix for Chapter 5
Independent Variables

I include a number of variables to control for the factors thought to influence the likelihood of nuclear acquisition. All control variables are drawn from Singh and Way, unless otherwise specified.[17] To assess a country's domestic capability to produce nuclear weapons, I include a measure of *GDP per capita*. To test for a non-monotonic relationship between level of economic development and nuclear acquisition, I include a squared term. *Industrial capacity index* is a dichotomous variable that measures whether a country produces steel domestically and has an electricity generating capacity greater than five thousand megawatts. States above a certain threshold of industrial development may be better able to support a nuclear weapons program.

To test the effect of a state's security environment on its risk of acquiring nuclear weapons, I include a *Rivalry* variable that measures whether a state is involved in at least one enduring rivalry.[18] The alternate *Disputes* variable is a five-year moving average of the number of militarized interstate disputes (MIDs) per year in which a state is involved.[19] *Alliance* is a dichotomous variable that assesses whether a state is in a defense pact with a nuclear-armed major power. States under the nuclear umbrella of an ally may have fewer incentives to develop their own nuclear capability. I also include an alternate *Superpower dependence* variable, indicating whether a state is in a defense pact with the United States or the Soviet Union. In chapter 1, it was argued that superpowers vigorously oppose nuclear proliferation and that states dependent on a superpower patron will be less likely to provide sensitive nuclear assistance. Following this logic, we may expect that states that are vulnerable to superpower pressure will also be less

---

2006; and Jasjeet S. Sekhon and Walter Mebane, "Genetic Optimization using Derivatives," *Political Analysis*, Vol. 7 (1998), pp. 187–210.

17. Singh and Way, "Correlates of Nuclear Proliferation."

18. Paul F. Diehl, *The Dynamics of Enduring Rivalries* (Urbana: University of Illinois Press, 1998); and Scott D. Bennett, "Integrating and Testing Models of Rivalry Termination," *American Journal of Political Science*, Vol. 42, No. 4 (October 1998), pp. 1200–1232.

19. Ghosen and Palmer, *Codebook for the Militarized Interstate Dispute Data*, Version 3.0, 2003.

likely to acquire nuclear weapons. Information on the coding of the superpower dependence variable can be found in chapter 2.

A number of variables gauge the institutional and economic determinants of nuclear acquisition. Scholars have argued that democratic states, due to their position in the "core" of the international system, may feel more secure and may be less likely to pursue nuclear weapons.[20] I include *Democracy,* which measures a country's domestic political regime type, drawing on data from the Polity IV index.[21] Similarly, Singh and Way have argued that "the spread of democracy reduces the likelihood that states will pursue nuclear weapons by enlarging a zone of peace."[22] *Percentage of democracies* is measured as the percentage of democratic states in the international system in each year. On the other hand, scholars have shown that periods of democratization are particularly unstable and encourage leaders to seek domestic support through the pursuit of nationalist policies, including possibly the pursuit of nuclear weapons.[23] To control for the relationship between regime change and nuclear acquisition, I include *Democratization,* which indicates a country's move toward democracy in three-, five-, and ten-year time spans. Scholars have also demonstrated that states that are open to the international economy or that are pursuing a strategy of economic liberalization are less likely to seek nuclear weapons because they are reluctant to risk international trade and investment on controversial foreign policies.[24] *Openness* assesses a state's openness to the international economy and is calculated as a country's trade ratio, exports plus imports, divided by GDP. *Liberalization* measures changes in a country's trade ratio over spans of three, five, and ten years.

Further Robustness Checks

I explore the robustness of my findings by examining the extent to which my results depend on the coding of the dependent variable and model specification. It is difficult to define precisely when some states acquired nuclear weapons. For states that have conducted a nuclear test, the date of nuclear acquisition is quite clear. For a nuclear weapon state that did not conduct a nuclear test, however, the date of the first assembly of nuclear weapons requires an examination of

---

20. See, for example, Glenn Chafetz, "The End of the Cold War and the Future of Nuclear Proliferation," in Davis and Frankel, eds., *Proliferation Puzzle.*
21. Jaggers and Gurr, "Tracking Democracy's Third Wave."
22. Singh and Way, "Correlates of Nuclear Proliferation," p. 864.
23. Edward D. Mansfield and Jack Snyder, "Democratization and the Danger of War," *International Security,* Vol. 20, No. 1 (Summer 1995), pp. 5–38; and Singh and Way, "Correlates of Nuclear Proliferation."
24. Solingen, "Political Economy of Nuclear Restraint"; Solingen, *Nuclear Logics;* and Paul, *Power Versus Prudence.*

the given country's historical record of nuclear development and some guesswork. There is also a question as to whether a country can enter the risk pool for nuclear acquisition multiple times. For example, India conducted its first nuclear explosion in 1974, but it is not believed to have assembled its first functioning nuclear weapon until 1988. Singh and Way code India as acquiring nuclear weapons in 1974, reentering the risk pool in 1975, and acquiring nuclear weapons a second time in 1988.[25] I code an alternate acquire variable in which states can only acquire nuclear weapons a single time, at which point they permanently exit the risk pool. Alternate codings of nuclear acquisition reveal that the results are not sensitive to various measurements or conceptualizations of the dependent variable.

Next, to ensure that my results were not being driven by the inclusion of specific control variables, I reran dozens of trimmed models that included different combinations of the subset of variables that were statistically significant in the above models. I also estimated simple bivariate models to assess the bivariate relationship between the key independent variables and nuclear proliferation. Again, the core results were not affected.

Appendix C: Cases of Sensitive Nuclear Assistance

U.S.S.R to China (1958–1960). The Soviet Union provided China with designs and key component parts for the Jiuquan plutonium reprocessing plant and for the Lanzhou uranium enrichment facility. Moscow reneged on a promise to provide Beijing with a prototype atomic bomb.[26]

France to Israel (1959–1965). France helped Israel to construct the Dimona plutonium reprocessing facility. The French are also believed to have transferred a nuclear weapon design.[27] French nuclear assistance was formally approved by the Guy Mollet government in 1956. According to a version of events propagated by investigative journalists Seymour Hersh, de Gaulle ordered a halt to all French assistance to Israel in 1960, but French bureaucracies continued sensitive nuclear cooperation with Tel Aviv without de Gaulle's knowledge or authorization.[28]

---

25. Singh and Way, "Correlates of Nuclear Proliferation."
26. Nuclear Threat Initiative, *Country Reports;* and Lewis and Xue, *China Builds the Bomb,* pp. 60–61, 72, 112, 118–121.
27. Nuclear Threat Initiative, *Country Reports;* Joseph Cirincione, with Jon B. Wolfsthal and Miriam Rajkumar, *Deadly Arsenals* (Washington, D.C.: Carnegie Endowment for International Peace, 2002), p. 225; Cohen, *Israel and the Bomb,* pp. 73–75; and Reed and Stillman, *Nuclear Express,* p. 79.
28. Hersh, *Samson Option.*

While Hersh's account makes for entertaining reading, the preponderance of evidence in other accounts indicate that de Gaulle made a calculated decision to cut off official government cooperation with Israel to provide the French government plausible deniability, while simultaneously using French firms to honor the nuclear cooperation agreement with Israel.[29]

France to Japan (1971–1974). France constructed a pilot-scale reprocessing plant for Japan at Tokai Mura.[30]

France to Pakistan (1974–1982). France helped Pakistan to construct the Chasma and Pinstech reprocessing plants. French assistance to the Chasma plant was halted in August 1978 under U.S. pressure. France continued the construction of the pilot-scale facility at Pinstech until its completion in 1982.[31]

France to Taiwan (1975). France agreed to provide Taiwan with a plutonium reprocessing facility. The French were able to transfer some of the component parts for the facility before Taiwan canceled the transaction under U.S. pressure. The United States dismantled the facilities related to reprocessing and confiscated the component parts.[32]

Italy to Iraq (1976–1978). Italy constructed a radiochemistry lab consisting of three lead-shielded hot cells capable of reprocessing plutonium in Iraq.[33]

Germany to Brazil (1979–1994). In 1975 Germany agreed to provide Brazil with ten nuclear reactors, a plutonium reprocessing plant, and a jet-nozzle uranium enrichment plant at Resende. Construction of the facilities began in 1979. After years of construction delays and cost overruns, Brazil decided to complete only two of the ten reactors and in 1985 indefinitely postponed the construction of the reprocessing plant. In March 1994, Brazil also canceled the construction of the uranium enrichment plant.[34]

France to Egypt (1980–1982). France constructed two hot cells for plutonium reprocessing in the Hot Laboratory and Waste Management Center in Egypt.[35]

---

29. See, for example, Peres, *Battling for Peace*, p. 142; Cohen, *Israel and the Bomb*, p. 75; and Reed and Stillman, *Nuclear Express*, pp. 80, 117.

30. Reiss, *Without the Bomb*, p. 115; Richard K. Lester, "U.S.-Japanese Nuclear Relations," *Asian Survey*, Vol. 22, No. 5 (May 1982), p. 422; and Robert W. Gale, "Nuclear Power and Japan's Proliferation Option," *Asian Survey*, Vol. 18, No. 11 (November 1978), p. 1124.

31. Weissman and Krosney, *Islamic Bomb*, pp. 74–84, 161–174; and Leonard S. Spector, *Nuclear Ambitions* (Boulder, CO: Westview Press, 1990), pp. 90–117.

32. Spector, *Nuclear Proliferation Today* (New York: Vintage, 1984), pp. 342–344; and Weissman and Krosney, *Islamic Bomb*, pp. 152–153.

33. Nuclear Threat Initiative, *Country Studies;* and Weissman and Krosney, *Islamic Bomb*, pp. 97–99.

34. Nuclear Threat Initiative, *Country Studies;* Jones et al., *Tracking Nuclear Proliferation*, pp. 231–242; and Spector, *Nuclear Ambitions*, pp. 242–266.

35. Nuclear Threat Initiative, *Country Studies;* Shyam Bhatia, *Nuclear Rivals in the Middle East* (New York: Routledge, 1988), p. 61.

China to Pakistan (1981–1983, 1984–1986). In the early 1980s, China supplied Pakistan with a nuclear weapon design and enough HEU for two nuclear weapons. Later, China is believed to have assisted Pakistan with the Kahuta uranium enrichment plant. In the 1990s, China also helped Pakistan with its reprocessing facility at Chasma and, in 1995, China exported five thousand ring magnets, component parts for uranium enrichment using the gaseous centrifuge method, to Pakistan.[36] Evan Medeiros claims that the 1995 ring magnet transfer (but not the nuclear assistance provided in the 1980s) may have been the result of lax export controls that allowed Chinese firms to export sensitive technology without the authorization of the central government.[37] China may have also conducted a joint nuclear test with Pakistan in 1990.[38] The nuclear cooperation in the 1990s occurred after Pakistan is believed to have achieved a nuclear weapons capability and are not recorded as cases of sensitive nuclear assistance.

China to Iran (1984, 1987, 1989, 1995). China provided Iran with calutrons, a key component part for uranium enrichment using the electromagnetic isotope separation method.[39]

China to Algeria (1986–1991). China constructed hot cells for Algeria at Ain Oussera and began the installation of a larger plutonium reprocessing facility.[40]

Pakistan to Iran (1987–1995). Pakistan provided Iran with designs and key component parts for a gaseous centrifuge uranium enrichment plant. It may have also transferred a nuclear weapon design.[41] These transfers were state-sponsored. In his memoirs, President Pervez Musharraf claims that the Pakistani government was unaware of, and did not authorize, sensitive nuclear exports to Iran, Libya, and North Korea.[42] This statement is belied by the overwhelming evidence that indicates that the nuclear exports were actively supported by senior government officials, including chiefs of army staff and civilian heads of state.[43]

---

36. Nuclear Threat Initiative, *Country Studies;* Jones et al., *Tracking Nuclear Proliferation,* pp. 50, 57–58n; Robert Shuey and Shirley A. Kan, "Chinese Missile and Nuclear Proliferation," *CRS Issue Brief* 29, September 9, 1995.
37. Evan S. Medeiros, *Chasing the Dragon* (Santa Monica, CA: RAND, 2005).
38. Reed and Stillman, *Nuclear Express,* pp. 131, 252–253.
39. Nuclear Threat Initiative, *Country Studies;* and Albright et al., *Plutonium and Highly Enriched Uranium* (Oxford: Oxford University Press, 1997), pp. 359–360.
40. David Albright and Corey Hinderstein, "Algeria: Big Deal in the Desert?" *Bulletin of the Atomic Scientists,* Vol. 57, No. 3 (May/June 2001), pp. 45–52; and Jones et al., *Tracking Nuclear Proliferation,* p. 163.
41. Nuclear Threat Initiative, *Country Studies;* Montgomery, "Ringing in Proliferation"; Corera, *Shopping for Bombs;* and Langewiesche, *Nuclear Bazaar.*
42. Musharraf, *In the Line of Fire.*
43. Bhatia, *Goodbye Shahzadi;* Corera, *Shopping for Bombs;* Frantz and Collins, *Nuclear Jihadist;* Langewiesche, *Nuclear Bazaar;* Levy and Scott-Clark, *Deception;* and Sagan, "Keeping the Bomb Away from Tehran," *Foreign Affairs,* Vol. 85, No. 5 (September/October 2006), p. 53.

Pakistan to Libya (1997–2001). Pakistan provided Libya with designs and key component parts for a gaseous centrifuge uranium enrichment plant. It also transferred a nuclear weapon design.[44]

Pakistan to North Korea (1997–2002). Pakistan provided North Korea with designs and key component parts for a gaseous centrifuge uranium enrichment plant. It may have also transferred a nuclear weapon design.[45]

Appendix D: Selected Cases of Nonsensitive Nuclear Assistance

Canada and the United States to India (1955). Canada supplied India with a nuclear reactor moderated with heavy water from the United States.[46] Nuclear reactors and heavy water do not qualify as sensitive nuclear assistance.

France to Japan (2001–present). The French firm AREVA assisted Japan in the construction of the Rokkasho-mura plutonium reprocessing facility.[47] This does not qualify as sensitive nuclear assistance because it does not materially advance Japan's ability to produce a nuclear weapon. Japan has enjoyed the ability to reprocess plutonium since 1977.[48]

Germany to Iraq (1985–1994). German firms exported materials that Iraq used in the construction of its nuclear facilities.[49] These materials consisted almost entirely of dual-use industrial materials, however, such as electrical components, industrial pipelines, soft iron, and furnace equipment. Germany did not export nuclear fuel-cycle facilities or their major component parts to Iraq.

Israel to South Africa (1977–1980). Israel may have provided missile technology and tritium to South Africa in exchange for natural uranium, but despite much suspicion, there is no evidence that Israel provided South Africa with sensitive nuclear assistance as I define it.[50] Some have speculated that South Africa may have also provided Israel with grounds for a nuclear test site, but this accusation has been dismissed by leading nuclear physicists.[51]

---

44. Ibid.
45. Ibid.
46. Perkovich, *India's Nuclear Bomb*.
47. AREVA, "AREVA to Pursue its Cooperation with JNFL to Start Up the Rokkasho-Mura Reprocessing Plant," December 20, 2005.
48. Reiss, *Without the Bomb*, p. 113.
49. Nuclear Threat Initiative, *Country Studies*; and Morstein and Perry, "Commercial Nuclear Trading Networks."
50. Liberman, "Rise and Fall."
51. Ruina, et al., "Ad hoc Panel Report on the September 22 Event."

The Netherlands to Pakistan (1974–1976). While working in the Netherlands in the mid-1970s, Pakistani scientist A. Q. Khan smuggled enrichment designs and equipment from the Netherlands to Pakistan without the knowledge and approval of the Dutch government.[52] This case does not qualify as sensitive nuclear assistance because it was not sponsored by the state.

North Korea or Pakistan to Libya (2000). In February 2005, the U.S. government charged North Korea with transferring uranium hexafluoride to Libya.[53] It is now believed that the uranium hexafluoride came from Pakistan.[54] Uranium hexafluoride does not qualify as sensitive nuclear assistance.

North Korea to Syria (2001–2007). North Korea helped Syria to construct a nuclear reactor.[55] The reactor was destroyed by Israel in a preventive military strike in September 2007. Nuclear reactors are not considered sensitive nuclear assistance. There is no evidence to suggest that North Korea also provided Syria with sensitive nuclear assistance.

Russia to Iran (1995–present). Russia rebuilt the Bushehr light-water nuclear power reactors in Iran, but nuclear reactors do not qualify as sensitive nuclear assistance. Russia considered constructing a uranium enrichment facility but canceled the deal under U.S. pressure.

Soviet Union to North Korea (1956–1967). The Soviet Union assisted North Korea with the construction of a research reactor and provided basic scientific training but did not assist North Korea with plutonium reprocessing or other sensitive nuclear technology.[56]

United States Atoms for Peace Program (1953–1975). Under the Atoms for Peace program initiated by President Dwight D. Eisenhower in 1953, the United States transferred research reactors and provided basic scientific training to many developing countries. The United States stopped well short, however, of providing sensitive nuclear assistance.[57]

United States to France (1970s and 1980s). During this period, the United States provided assistance to France designed primarily to improve the safety and security of French warheads.[58] This does not qualify as sensitive nuclear as-

---

52. Corera, *Shopping for Bombs.*
53. See, for example, Glenn Kessler, "North Korea May Have Sent Libya Nuclear Material, U.S. Tells allies," *Washington Post,* February 2, 2005, p. A01.
54. Dafna Linzer, "U.S. Misled Allies About Nuclear Export: North Korea Sent Material to Pakistan, Not Libya," *New York Times,* March 20, 2005, A01.
55. Paul Richter and Greg Miller, "U.S. Offers Evidence of North Korea-Syria Nuclear Plant," *Los Angeles Times,* April 25, 2008.
56. Wit et al., *Going Nuclear.*
57. Hewlett and Holl, *Atoms for Peace and War.*
58. Richard H., Ullman, "The Covert French Connection," *Foreign Policy,* No. 75 (Summer 1989), pp. 3–33.

sistance because France was, at this time, already an established nuclear weapon state. The United States did not provide sensitive nuclear assistance—indeed, Washington actively denied sensitive nuclear transfers—to France before Paris acquired the bomb.[59]

United States to India (1961, 2005). A U.S. firm, Vitro International, prepared blueprints for the construction of the physical site at the Trombay reprocessing facility in 1961 but did not work on the sensitive nuclear technologies. The sensitive technologies in the Trombay reprocessing facility were developed autonomously with the aid of declassified documents on plutonium reprocessing made available by the UN.[60] The U.S.-India nuclear deal signed in January 2005 is related to civilian nuclear assistance only and does not provide for the transfer of sensitive nuclear materials or technology.

Appendix E: Selected Cases of Nonassistance

Argentina to Iran (1992). Argentina denied an Iranian request for hot cells.[61]

Argentina to Libya (1985). Argentina offered to sell reprocessing technology to Libya but canceled the deal under U.S. pressure.[62]

China, Germany, Great Britain, and Yugoslavia to Iraq (1987–1990). Iraq was able to acquire component parts and materials to be used in its uranium enrichment program from various suppliers in Europe and Asia. These were piecemeal acquisitions, however, consisting almost entirely of unrestricted dual-use items. Iraq never received a substantial infusion of sensitive nuclear assistance from abroad in this time period.[63]

China to North Korea (1964). China denied a North Korean request for assistance with nuclear weapons technology.[64]

France to South Korea (1975–1976). France agreed to sell reprocessing technology to South Korea but canceled the deal under U.S. pressure.[65]

---

59. See, for example, Reed and Stillman, *Nuclear Express*, p. 116.
60. Perkovich, *India's Nuclear Bomb*, p. 28; Jones et al., *Tracking Nuclear Proliferation*, p. 112; Reed and Stillman, *Nuclear Express*, p. 159.
61. Mark Hibbs, "Iran Sought Sensitive Nuclear Supplies from Argentina, China," *Nucleonics Week,* September 24, 1992, pp. 2–3.
62. Jones et al., *Tracking Nuclear Proliferation*, p. 224.
63. David Albright and Mark Hibbs, "Iraq's Shop-til-You-Drop Nuclear Program," *Bulletin of Atomic Scientists*, Vol. 48, No. 3 (April 1992), pp. 26–37.
64. Wit et al., *Going Critical.*
65. Katz et al., eds., *Nuclear Power in Developing Countries*, p. 227.

France to Australia (1969). In 1969 France agreed to help Australia develop uranium enrichment capabilities, but the deal was cancelled after initial scientific visits and before any sensitive nuclear transfers took place.[66]

Germany to South Africa (1968–1972). There has been speculation, but no concrete evidence, that a German firm assisted South Africa with jet-nozzle uranium enrichment after the German cabinet explicitly prohibited the cooperation.[67] If the cooperation did occur, it was not state-sponsored and therefore would not count as sensitive nuclear assistance as I define it.

Italy to Argentina (1969). Experts once speculated that Italy may have assisted Argentina in the construction of the Ezeiza reprocessing facility. It is now believed that the facility was constructed indigenously.[68]

Norway to Yugoslavia (1966). Norway considered selling reprocessing technology to Yugoslavia, but the deal was never consummated.[69]

Pakistan to Iraq and Syria (1990). Pakistan may have offered Iraq and Syria uranium enrichment technology, but the deals were never consummated.[70]

United States to Great Britain (1940–1951 and 1960s). Contrary to the belief of many, the United States did not directly assist the British nuclear weapons program. Rather, during the Manhattan Project, the United States selectively exploited Britain's expertise in nuclear physics while systematically denying the British access to America's most sensitive nuclear research. Official U.S. policy in this period was to prevent Britain from obtaining the bomb.[71] The United States assisted Britain with strategic delivery vehicles in the 1960s, but this does not qualify as sensitive nuclear assistance for two reasons. First, by this time, Great Britain was already an established nuclear weapon state. Second, delivery vehicles are not considered sensitive nuclear assistance.

---

66. Hymans, *Psychology of Nuclear Proliferation*, p. 132.
67. Barbara Rogers and Cervenka Zedenk, *Nuclear Axis* (New York: Times Books, 1978).
68. Spector, *Nuclear Proliferation Today*, p. 203.
69. William C. Potter, Djuro Miljanic, and Ivo Slaus, "Tito's Nuclear Legacy," *Bulletin of Atomic Scientists*, Vol. 56, No. 2 (March/April 2000), pp. 63–70.
70. Montgomery, "Ringing in Proliferation," p. 173.
71. See, for example, Gowing, *Britain and Atomic Energy*; Gowing, *Independence and Deterrence*; Richard Rhodes, *The Making of the Atomic Bomb* (New York: Simon and Schuster, 1995); and Reed and Stillman, *Nuclear Express*, pp. 26–53.

Bibliography

"The A. Q. Khan Network: Case Closed?" Hearing before the Subcommittee on International Terrorism and Nonproliferation of the Committee on International Relations, House of Representatives, May 25, 2006.

Abbas, Hassan, *Pakistan's Drift into Extremism: Allah, the Army, and America's War on Terror* (New York: M.E. Sharpe, 2004).

Abernathy, David B., *The Dynamics of Global Dominance: European Overseas Empires, 1415–1980* (New Haven: Yale University Press, 2002).

Abraham, Itty, "The Ambivalence of Nuclear Histories," *Osiris*, Vol. 21 (2006), pp. 49–65.

———, *The Making of the Indian Atomic Bomb: Science, Society, and the Postcolonial State* (London: Zed Books, 1998).

Aburish, Said K., *Nasser: The Last Arab* (London: Thomas Dunne Books, 2004).

Accinelli, Robert, *Crisis and Commitment: United States Policy toward Taiwan, 1950–1955* (Chapel Hill: University of North Carolina Press, 1996).

Achen, Christopher, and Duncan Snidal, "The Rational Deterrence Debate: A Symposium Rational Deterrence Theory and Comparative Case Studies," *World Politics*, Vol. 41, No. 2 (January 1989), pp. 143–169.

Adams, James, *Israel and South Africa: The Unnatural Alliance* (London: Quartet Books, 1984).

Ahmed, Samina, "Pakistan's Nuclear Weapons Program," *International Security*, Vol. 23, No. 4 (Spring 1999), pp. 178–204.

Albright, David, "An Iranian Bomb," *Bulletin of the Atomic Scientists*, Vol. 51, No. 4 (July/August 1995), pp. 20–26.

———, "South Africa's Secret Nuclear Weapons," *ISIS Report*, Vol. 1, No. 4 (Washington, D.C.: Institute for Science and International Security, May 1994), pp. 6–8.

Albright, David, Frans Berkhout, and William Walker, *Plutonium and Highly Enriched Uranium 1996: World Inventories, Capabilities, and Policies* (Oxford: Oxford University Press, 1997).

Albright, David, and Mark Hibbs, "Iraq's Shop-til-You-Drop Nuclear Program," *Bulletin of Atomic Scientists*, Vol. 48, No. 3 (April 1992), pp. 26–37.

———, "Pakistan's Bomb: Out of the Closet," *Bulletin of Atomic Scientists*, Vol. 48, No. 6 (July/August 1992), pp. 38–43.

Albright, David, and Corey Hinderstein, "Algeria: Big Deal in the Desert?" *Bulletin of the Atomic Scientists*, Vol. 57, No. 3 (May/June 2001), pp. 45–52.

Alexander, Martin S., *France and the Algerian War, 1954–1962: Strategy, Operations, and Diplomacy* (New York: Routledge, 2002).

Alexander, Martin S., Martin Evans, and J. F. V. Keigler, eds., *The Algerian War and the French Army, 1954–1962: Experiences, Images, Testimonies* (New York: Palgrave Macmillan, 2002).

Allison, Graham, "Deterring Kim Jong Il," *Washington Post*, October 27, 2006, p. A23.

———, "Making Good on Bush's Vow Will Require Detective Work," *New York Times*, October 13, 2006, p. A12.

———, *Nuclear Terrorism: The Ultimate Preventable Catastrophe* (New York: Times Books, 2005).

Allon, Yigal, *Contriving Warfare* (in Hebrew) (Tel Aviv: Hakibuttz Hameuchad, 1990).
———, *Curtain of Sand* (in Hebrew) (Tel Aviv: Hakibuttz Hameuchad, 1968).
Alteras, Isaac, *Eisenhower and Israel: U.S.-Israeli Relations, 1953–1960* (Gainesville: University Press of Florida, 1993).
Alterman, Jon B., *Egypt and American Foreign Assistance, 1952–1956: Hopes Dashed* (New York: Palgrave Macmillan, 2002).
AREVA, "AREVA to Pursue its Cooperation with JNFL to Start Up the Rokkasho-Mura Reprocessing Plant," December 20, 2005, available at www.areva.com.
Atomic Energy: Cooperation for Civil Use, Agreement between the United States of America and Israel (Treaties and Other International Acts Series 3311) (Washington, D.C.: Department of State, publication 5963).
Baev, Pavel K., "Deal with Tehran Extends Russia's Dwindling Sphere of Influence: Moscow Insists on Seeing No Evil in Iran," *Eurasia Daily Monitor*, Vol. 2, No. 43 (December 3, 2005).
Bar-Zohar, Michael, *Ben Gurion* (Tel Aviv: Smora Bitan, 1987).
Barbash, Fred, "Iran Says It Would Transfer Nuclear Technology," *Washington Post*, April 26, 2006, p. A19.
Barker, A. J., *Suez: The Seven Day War* (London: Faber and Faber, 1964).
Barletta, Michael, "Democratic Security and Diversionary Peace: Nuclear Confidence-Building in Argentina and Brazil," *National Security Studies Quarterly*, Vol. 5, No. 3 (Summer 1999), pp. 19–38.
Barnes, Jonathan, *Cambridge Companion to Aristotle* (Cambridge: Cambridge University Press, 1995).
Barrett, Roby C., *The Greater Middle East and the Cold War: U.S. Foreign Policy under Eisenhower and Kennedy* (New York: I.B. Tauris, 2004).
Bass, Warren, *Support Any Friend: Kennedy's Middle East and the Making of the U.S.-Israel Alliance* (Oxford: Oxford University Press, 2003).
Beardsley, Kyle, and Victor Asal, "Winning with the Bomb," *Journal of Conflict Resolution*, Vol. 53, No. 2 (April 2009).
Beck, Nathaniel, Jonathan N. Katz, and Richard Tucker, "Taking Time Seriously in Binary Time-Series—Cross-Section Analysis," *American Journal of Political Science*, Vol. 42, No. 4 (October 1998), pp. 1260–1288.
"Ben Gurion Explains Nuclear Project," *New York Times*, December 19, 1960, p. 8.
Bennett, Scott D., "Integrating and Testing Models of Rivalry Termination," *American Journal of Political Science*, Vol. 42, No. 4 (October 1998), pp. 1200–1232.
Bennett, Scott D., and Allan Stam, *The Behavioral Origins of War* (Ann Arbor: University of Michigan Press, 2003).
———, "*EUGene*: A Conceptual Manual." *International Interactions*, Vol. 26, No. 2 (2000), pp. 179–204.
Berman, Ilan, "Confronting a Nuclear Iran," Testimony before the U.S. House of Representatives, Committee on Armed Services, February 1, 2006.
Betts, Richard K., *Nuclear Blackmail and Nuclear Deterrence* (Washington, D.C.: Brookings Institution Press, 1987).
———, "Paranoids, Pygmies, Pariahs, and Nonproliferation Revisited," in Zachary S. Davis and Benjamin Frankel, eds., *The Proliferation Puzzle: Why Nuclear Weapons Spread (and What Results)* (Portland: Frank Cass, 1993).
———, "Systems for Peace or Causes of War? Collective Security, Arms Control and the New Europe," *International Security*, Vol. 17, No. 1 (Summer 1992), pp. 5–43.
———, "Universal Deterrence or Conceptual Collapse? Liberal Pessimism and Utopian Realism," in Victor A. Utgoff, ed., *The Coming Crisis: Nuclear Proliferation, U.S. Interests, and World Order* (Cambridge, MA: MIT Press, 2000).

Bhatia, Shyam, *Goodbye Shahzadi* (New Delhi: Eastern Book Corporation, 2008).
———, *Nuclear Rivals in the Middle East* (New York: Routledge, 1988).
Biddle, Stephen, "Afghanistan and the Future of Warfare," *Foreign Affairs*, Vol. 82, No. 2 (March/April 2003).
———, *Military Power: Explaining Victory and Defeat in Modern Battle* (Princeton, NJ: Princeton University Press, 2004).
Bissell, Richard M., Jr., Jonathan E. Lewis, and Frances T. Pudlow, *Reflections of a Cold Warrior: From Yalta to the Bay of Pigs* (New Haven: Yale University Press, 1996).
Blair, Bruce G., *The Logic of Accidental Nuclear War* (Washington, D.C.: Brookings University Press, 1993).
———, "Nuclear Inadvertence: Theory and Evidence," *Security Studies*, Vol. 3, No. 3 (Spring 1994), pp. 494–500.
Blair, Dennis C., "Annual Threat Assessment of the Intelligence Community for the Senate Select Committee on Intelligence," February 12, 2009, unclassified Statement for the Record.
Blanton, Shannon Lindsey, "Foreign Policy in Transition? Human Rights, Democracy, and U.S. Arms Exports," *International Studies Quarterly*, Vol. 49, No. 4 (December 2005), pp. 647–667.
———, "Promoting Human Rights and Democracy in the Developing World: U.S. Rhetoric versus U.S. Arms Exports," *American Journal of Political Science*, Vol. 44, No. 1 (January 2000), pp. 123–131.
Boulding, Kenneth, *Conflict and Defense* (New York: Harper and Brothers, 1962).
Box-Steffensmeier, Janet M., and Bradford S. Jones, "Time Is of the Essence: Event History Models in Political Science," *American Journal of Political Science*, Vol. 41 (1997), pp. 1414–1461.
Boyne, Walter J., *The Two O'Clock War: The 1973 Yom Kippur Conflict and the Airlift that Saved Israel* (New York: Thomas Dunne Books, 2002).
Bozoft, Farzad, "Iran Signs Secret Atom Deal," *The Observer*, June 12, 1988.
Braun, Chaim, and Christopher Cyba, "Proliferation Rings: New Challenges to the Nuclear Nonproliferation Regime," *International Security*, Vol. 29, No. 2 (Fall 2004), pp. 5–49.
Brecher, Michael, *Decisions in Israel's Foreign Policy* (New Haven: Yale University Press, 1975).
Brodie, Bernard, *The Absolute Weapon: Atomic Power and World Order* (Manchester, NH: Ayer, 1946).
———, *Strategy in the Missile Age* (Princeton, NJ: Princeton University Press, 1959).
Brookes, Peter, "Iran Emboldened: Tehran Seeks to Dominate Middle East Politics," *Armed Forces Journal*, April 2, 2007.
Bruce St. John, Ronald, *Libya and the United States: Two Centuries of Strife* (Philadelphia: University of Pennsylvania Press, 2002).
Brugioni, Dino A., *Eyeball to Eyeball: The Inside Story of the Cuban Missile Crisis* (New York: Random House, 1991).
Bueno de Mesquita, Bruce, *The War Trap* (New Haven: Yale University Press, 1981).
Bueno de Mesquita, Bruce, and William. H. Riker, "An Assessment of the Merits of Selective Nuclear Proliferation," *Journal of Conflict Resolution*, Vol. 26, No. 2 (June 1982), pp. 283–306.
Bujon de L'Estang, Francois, "The Delicate Balance: Government and Industry Cooperation in Enforcing Nonproliferation," in Rodney W. Jones, Cesare Merlini, Joseph F. Pilat, and William C. Potter, eds., *The Nuclear Suppliers and Nonproliferation: International Policy Choices* (Lexington, MA: Lexington Books, 1985).
Bundy, McGeorge, *Danger and Survival: Choices about the Bomb in the First Fifty Years* (New York: Random House, 1988).

Bunn, Matthew, and Anthony Wier, "Terrorist Nuclear Weapon Construction: How Difficult?" *Annals of the American Academy of Political and Social Science* (2006).
Burr, William, and Jeffrey T. Richelson, "A Chinese Puzzle," *Bulletin of the Atomic Scientists*, Vol. 53, No. 4 (July/August 1997), pp. 42–47.
———, "Whether to Strangle the Baby in the Cradle: The United States and the Chinese Nuclear Program, 1960–64," *International Security*, Vol. 25, No. 3 (Winter 2000), pp. 54–99.
Calvin, James Barnard, *The China-India Border War (1962)* (Marine Corps Command and Staff College, April 1984).
Carr, E. H., *The Twenty Years Crisis: 1919–1939* (New York: Macmillan, 1939).
Central Intelligence Agency, Office of National Estimate, Memorandum for the Director, Sherman Kent, "Consequences of Israeli Acquisition of Nuclear Capability," March 6, 1963, 1, NSF, Box 118, John F. Kennedy Library.
Cerny, Philip G., *The Politics of Grandeur: Ideological Aspects of de Gaulle's Foreign Policy* (Cambridge: Cambridge University Press, 1980).
Cha, Victor D., and David C. Kang, *Nuclear North Korea: A Debate on Engagement Strategies* (New York: Columbia University Press, 2003).
Chafetz, Glenn, "The End of the Cold War and the Future of Nuclear Proliferation: An Alternative to the Neorealist Perspective," in Zachary S. Davis and Benjamin Frankel, eds., *The Proliferation Puzzle: Why Nuclear Weapons Spread (and What Results)* (Portland: Frank Cass, 1993).
Chestnut, Sheena, "Illicit Activity and Proliferation: North Korean Smuggling Networks," *International Security*, Vol. 32, No. 1 (Summer 2007), pp. 80–111.
Chivers, Daniel H., Bethany F. Lyles, Brett H. Isselhardt, and Jonathan S. Snyder, "Before the Day After: Using Pre-Detonation Nuclear Forensics to Improve Fissile Material Security," *Arms Control Today* (July/August 2008), available at www.arms control.org/act/2008_07-08/NuclearForensics.
Christofferson, Thomas R., with Michael S. Christofferson, *France during World War II: From Defeat to Liberation* (New York: Fordham University Press, 2006).
CIA World Factbook (Washington, D.C.: Central Intelligence Agency, 2008).
Cirincione, Joseph, *Bomb Scare: The History and Future of Nuclear Weapons* (New York: Columbia University Press, 2007).
Cirincione, Joseph, with Jon B. Wolfsthal and Miriam Rajkumar, *Deadly Arsenals: Tracking Weapons of Mass Destruction* (Washington, D.C.: Carnegie Endowment for International Peace, 2002).
Coe, Andrew, J., "North Korea's New Cash Crop," *Washington Quarterly*, Vol. 28, No. 3 (Summer 2005), pp. 73–84.
Cohen, Avner, *Israel and the Bomb* (New York: Columbia University Press, 1998).
Cointet, Jean-Paul, "Guy Mollet, the French Government, and the SFIO," in Selwyn Ilan Troen and Moshe Shemesh, eds., *The Suez-Sinai Crisis 1956: Retrospective and Reappraisal* (New York: Columbia University Press, 1990).
Conze, Henri, former official, French Ministry of Defense, interview with the author, Paris, France, June 2007.
Cordesman, Anthony H., *Lessons of Modern War: The Iran-Iraq War* (London: Mansell Publishing, 1990).
Corera, Gordon, *Shopping for Bombs: Nuclear Proliferation, Global Insecurity, and the Rise and Fall of the A. Q. Khan Network* (Oxford: Oxford University Press, 2006).
Correlates of War Project, *National Material Capabilities Data Documentation*, version 3.0, May 2005.
Crile, George, *Charlie Wilson's War: The Extraordinary Story of the Largest Covert Operation in History* (New York: Atlantic Monthly Press, 2003).

Crosbie, Sylvia, *A Tacit Alliance: France and Israel from Suez to the Six-Day War* (Princeton, NJ: Princeton University Press, 1974).
Dallek, Robert, *Nixon and Kissinger: Partners in Power* (New York: HarperCollins, 2007).
Danilovic, Vesna, "The Sources of Threat Credibility in Extended Deterrence," *Journal of Conflict Resolution*, Vol. 45, No. 3 (2001), pp. 341–369.
Davies, Peter, *France and the Second World War* (New York: Taylor and Francis, 2007).
Davis, Paul K., and Brian M. Jenkins, *Deterrence and Influence in Counterterrorism: A Component in the War on al Qaeda* (Santa Monica, CA: RAND, 2002).
Davis, Zachary S., "The Realist Nuclear Regime," in Zachary S. Davis and Benjamin Frankel, eds., *The Proliferation Puzzle: Why Nuclear Weapons Spread (and What Results)* (Portland: Frank Cass, 1993).
Dawisha, Karen, *Soviet Foreign Policy toward Egypt* (New York: St. Martin's, 1979).
The Decommissioning of the Eurochemic Reprocessing Plant, available at www.belgoprocess.be/03_act/docs/BP02_Eurochemic.pdf.
Deutsch, Karl W., and David J. Singer, "Multipolar Power Systems and International Stability," *World Politics*, Vol. 16, No. 3 (April 1964), pp. 390–406.
Diehl, Paul F., *The Dynamics of Enduring Rivalries* (Urbana: University of Illinois Press, 1998).
Dixit, J. N., *India-Pakistan in War and Peace* (New York: Taylor and Francis, 2007).
Donaldson, Robert H., and John A. Donaldson, "The Arms Trade in Russian-Chinese Relations: Identity, Domestic Politics, and Geopolitical Positioning," *International Studies Quarterly*, Vol. 47, No. 4 (December 2003), pp. 709–732.
Dong, Wonmo, *The Two Koreas and the United States: Issues of Peace, Security, and Economic Cooperation* (New York: East Gate Book, 2000).
Downes, Alexander, "Desperate Times, Desperate Measures. The Causes of Civilian Victimization in War," *International Security*, Vol. 30, No. 4 (Spring 2006), pp. 152–195.
———, *Targeting Civilians in War* (Ithaca, NY: Cornell University Press, 2008).
Doyle, Michael W., "Liberalism and World Politics," *American Political Science Review*, Vol. 80, No. 4 (December 1986), pp. 1151–1169.
———, *Ways of War and Peace: Realism, Liberalism, and Socialism* (New York: W.W. Norton, 1997).
Drell, Sidney D., and James E. Goodby, *The Gravest Danger: Nuclear Weapons* (Stanford, CA: Hoover Institution Press, 2003).
Drucks, Herbert, *The Uncertain Alliance: The U.S. and Israel from Kennedy to the Peace Process* (New York: Greenwood Press, 2001).
Dugger, Celia W., "The Kashmir Brink," *New York Times*, June 20, 2002.
Dunlop, William, and Harold Smith, "Who Did It? Using International Forensics to Detect and Deter International Terrorism," *Arms Control Today*, Vol. 36, No. 8 (October 2006).
Dunn, Lewis, "Can Al Qaeda Be Deterred from Using Nuclear Weapons?" *Center for the Study of Weapons of Mass Destruction Occasional Paper*, No. 3 (Washington, D.C.: National Defense University Press, July 2005).
Dunn, Lewis, Matthew Kroenig, Harold Smith, and Steven Weber, *International Ramifications of Nuclear Terrorism*, prepared for the Defense Threat Reduction Agency (August 2005).
Dunnigan, James F., *How to Make War: A Comprehensive Guide to Warfare in the Twenty-First Century* (New York: Harper, 2003) 4th ed.
Dunstan, Simon, *The Yom Kippur War: The Arab-Israeli War of 1973* (New York: Osprey Publishing, 2007).
El-Baradei, Mohamed, "Towards a Safer World," *Economist*, October 16, 2003.

Elworthy, Scilla, *How Nuclear Weapons Decisions Are Made* (New York: St. Martin's, 1986).
Fayazmanesh, Sas, *The United States and Iran: Sanctions, Wars, and the Policy of Dual Containment* (New York: Routledge, 2008).
Fearon, James D., "Domestic Political Audiences and the Escalation of International Disputes," *American Political Science Review*, Vol. 88, No. 3 (September 1994), pp. 577–592.
Feaver, Peter Douglas, *Guarding the Guardians: Civilian Control of Nuclear Weapons in the United States* (Ithaca, NY: Cornell University Press, 1993).
——, "The Politics of Inadvertence," *Security Studies*, Vol. 3, No. 3 (Spring 1994), pp. 501–508.
Feaver, Peter Douglas, and Emerson M. S. Niou, "Managing Nuclear Proliferation: Condemn, Strike, or Assist," *International Studies Quarterly*, Vol. 40, No. 2 (Summer 1996), pp. 209–233.
Ferguson, Charles D., *Nuclear Energy: Balancing Benefits and Risks*, Council on Foreign Relations, Special Report 28 (2007).
Fetzer, James, "Clinging to Containment: China Policy," in Thomas G. Paterson, ed., *Kennedy's Quest for Victory: American Foreign Policy, 1961–63* (New York: Oxford University Press, 1989).
Finney, John W., "U.S. Hears Israel Moves Toward A-Bomb Potential," *New York Times*, December 19, 1960, p. 1.
Fischer, David, "South Africa," in Mitchell Reiss and Robert S. Litwak, eds., *Nuclear Proliferation after the Cold War* (Washington, D.C.: Woodrow Wilson Center Press, 1994).
"France Admits It Gave Israel A-Bomb," *Sunday Times* (London), October 12, 1986.
Frankel, Benjamin, "The Brooding Shadow: Systemic Incentives and Nuclear Weapons Proliferation," in Zachary S. Davis and Benjamin Frankel, eds., *The Proliferation Puzzle: Why Nuclear Weapons Spread (and What Results)* (Portland: Frank Cass, 1993).
Frantz, Douglas, and Kathleen Collins, *The Nuclear Jihadist: The True Story of the Man Who Sold the World's Most Dangerous Secrets...and How We Could Have Stopped Him* (New York: Twelve, 2007).
Fravel, M. Taylor, "Power Shifts and Escalation: Explaining China's Use of Force in Territorial Disputes," *International Security*, Vol. 32, No. 3 (Winter 2007/2008), pp. 44–83.
Freeman, Colin, and Philip Sherwell, "Iranian Fatwa Approves Use of Nuclear Weapons," *The Telegraph* (London), February 19, 2006.
Fuhrmann, Matthew, "Exporting Mass Destruction? The Determinants of Dual-Use Trade," *Journal of Peace Research*, Vol. 45, No. 5 (September 2008).
——, "The Nuclear Marketplace and Grand Strategy," Ph.D. dissertation, University of Georgia, 2008.
——, "Taking a Walk on the Supply Side: The Determinants of Civilian Nuclear Cooperation," *Journal of Conflict Resolution*, Vol. 53, No. 2 (April 2009).
Gaddis, John Lewis, *The Cold War: A New History* (New York: Penguin, 2005).
——, *We Now Know: Rethinking Cold War History* (Oxford: Oxford University Press, 1997).
Gaddy, Clifford G., *The Price of the Past: Russia's Struggle with the Legacy of a Militarized Economy* (Washington, D.C.: Brookings Institution Press, 1998).
Gale, Robert W., "Nuclear Power and Japan's Proliferation Option," *Asian Survey*, Vol. 18, No. 11 (November 1978), pp. 1117–1133.
Gall, Norman, "Atoms for Brazil, Dangers for All," *Foreign Policy*, Vol. 23 (Summer 1976), pp. 155–201.
Gallucci, Robert L., "Averting Nuclear Catastrophe," *Harvard International Review*, Vol. 26, No. 4 (Winter 2005), available at http://hir.harvard.edu/articles/1303/.

Ganguly, Sumit, "India's Pathway to Pokhran II." *International Security,* Vol. 23, No. 4 (Spring 1999), pp. 148–177.
Garrett, Geoffrey, and Barry Weingast, "Ideas, Interests, and Institutions: Constructing the European Community's Internal Market," in Judith Goldstein and Robert Keohane, eds., *Ideas and Foreign Policy: Beliefs, Institutions, and Political Change* (Ithaca, NY: Cornell University Press, 1993).
Garthoff, Raymond, *Détente and Confrontation: American Soviet Relations from Nixon to Reagan* (Washington, D.C.: Brookings Institution, 1994).
Gartzke, Erik, *Codebook for the Affinity of Nations Index, 1946–2002,* version 3.0, 2006.
Gartzke, Erik, and Dong-Joon Jo, "Bargaining, Nuclear Proliferation, and Interstate Disputes," *Journal of Conflict Resolution,* Vol. 53, No. 2 (April 2009).
Gartzke, Erik, and Matthew Kroenig, "A Strategic Approach to Nuclear Proliferation," *Journal of Conflict Resolution,* Vol. 53, No. 2 (April 2009).
Garver, John W., *Sino-Indian Rivalry in the Twenty-First Century* (Seattle: University of Washington Press, 2001).
Gavin, Francis. J., "Blasts from the Past: Proliferation Lessons from the 1960s," *International Security,* Vol. 29, No. 1 (Winter 2004), pp. 100–135.
Geddes, Barbara, "How the Cases You Choose Affect the Answers You Get: Selection Bias in Comparative Politics," *Political Analysis,* Vol. 2, No. 1 (1990), pp. 131–150.
George, Alexander L., "Case Studies and Theory Development: The Method of Structured, Focused Comparison," in Paul Gordon Lauren, ed., *Diplomacy: New Approaches in History, Theory, and Policy* (New York: Free Press, 1979).
George, Alexander L., and Richard Smoke, *Deterrence in American Foreign Policy: Theory and Practice* (New York: Columbia University Press, 1974).
Gertz, Bill, "CIA Report Cites N. Korean Proliferation Threat," *Washington Times,* November, 27 2004, accessed at www.washingtontimes.com/national/20041126-111219-4624r.htm.
Ghosen, Faten, and Glenn Palmer, *Codebook for the Militarized Interstate Dispute Data,* Version 3.0, 2003, accessed from http://cow2.la.psu.edu.
Gibler, Douglas M., and Meredith Sarkees, *Coding Manual for v3.0 of the Correlates of War Formal Interstate Alliance Data Set, 1816–2000,* 2002, unpublished manuscript.
Gilpin, Robert, *War and Change in World Politics* (Cambridge: Cambridge University Press 1981).
Ginor, Isabella, and Gideon Remez, *Foxbats over Dimona: The Soviets' Nuclear Gamble in the Six-Day War* (New Haven: Yale University Press, 2007).
Glaser, Charles L., *Analyzing Strategic Nuclear Policy* (Princeton, NJ: Princeton University Press, 1991).
Glaser, Charles L., and Steve Fetter, "National Missile Defense and the Future of U.S. Nuclear Weapons Policy," *International Security,* Vol. 26, No. 1 (Summer 2001), pp. 40–92.
Gleditsch, Kristian, "Expanded Trade and GDP Data," *Journal of Conflict Resolution,* Vol. 46, No. 5 (2002), pp. 712–724.
Goertz, Gary, and Paul F. Diehl, "Enduring Rivalries: Theoretical Constructs and Empirical Patterns, *International Studies Quarterly,* Vol. 37, No. 2 (1993), pp. 147–172.
——, "The Initiation and Termination of Enduring Rivalries: The Impact of Political Shocks," *American Journal of Political Science,* Vol. 39, No. 1 (1995), pp. 30–52.
Goldstein, Lyle J., *Preventive Attack and Weapons of Mass Destruction: A Comparative Historical Analysis* (Stanford, CA: Stanford University Press, 2006).
Goncharenko, Sergei, "Sino-Soviet Military Cooperation," in Odd Arne Westad, ed., *Brothers in Arms: The Rise and Fall of the Sino-Soviet Alliance, 1945–1963* (Washington, D.C.: Woodrow Wilson Center Press, 1998).

Gorst, Anthony, and Lewis Johnman, *The Suez Crisis* (London: Routledge, 1997).
Gourevitch, Peter, *Politics in Hard Times: Comparative Responses to International Economic Crises* (Ithaca, NY: Cornell University Press, 1986).
Government Accountability Office Report, *Nuclear Nonproliferation: U.S. Efforts to Help Other Countries Combat Nuclear Smuggling Need Strengthened Coordination and Planning*, May 2002.
Gowa, Joanne, "Rational Hegemons, Excludable Goods, and Small Groups: An Epitaph for Hegemonic Stability Theory?" *World Politics*, Vol. 41, No. 3 (1989), pp. 307–324.
——, *Allies, Adversaries, and International Trade* (Princeton, NJ: Princeton University Press, 1994).
Gowing, Margaret, *Britain and Atomic Energy, 1939–1945* (London: Macmillan, 1964).
——, *Independence and Deterrence: Britain and Atomic Energy 1945–1952*, Vols. 1 and 2 (London: Macmillan, 1974).
Government Accountability Office, *Nuclear Nonproliferation: U.S. Efforts to Help Other Countries Combat Nuclear Smuggling Need Strengthened Coordination and Planning*, May 2002.
Gregory, Barbara M., "Egypt's Nuclear Program: Assessing Supplier Based and Other Developmental Constraints," *Nonproliferation Review*, Vol. 3, No. 1 (Fall 1995), pp. 20–27.
Grieco, Joseph, "The Relative Gains Problem for International Cooperation," *American Political Science Review*, Vol. 87, No. 3 (September 1993), pp. 729–735.
Gromyko, Andrei A., *Memoirs* (New York: Doubleday, 1989), pp. 251–252.
Hagerty, Devin T., *The Consequences of Nuclear Proliferation: Lessons from South Asia* (Cambridge, MA: MIT Press, 1998).
Haggard, Stephen, and Beth A. Simmons, "Theories of International Regimes," *International Organization*, Vol. 41, No. 3 (Summer 1987), pp. 491–517.
Halperin, Morton, *Limited War in the Nuclear Age* (Santa Monica, CA: RAND, 1963).
Hamza, Khidhir, "Inside Saddam's Secret Nuclear Program," *Bulletin of the Atomic Scientists*, Vol. 54, No. 5 (September/October 1998), pp. 26–33.
Haqqani, Husain, *Pakistan: Between Mosque and Military* (Washington, D.C.: Carnegie Endowment for International Peace, 2005).
Hersh, Seymour, *The Price of Power: Kissinger in the Nixon White House* (New York: Summit, 1983).
——, *The Samson Option: Israel's Nuclear Arsenal and American Foreign Policy* (New York: Random House, 1991).
Hewlett, Richard G., and Jack M. Holl, *Atoms for Peace and War, 1953–1961: Eisenhower and the Atomic Energy Commission* (Berkeley: University of California Press, 1989).
Hibbs, Mark, "Iran Sought Sensitive Nuclear Supplies from Argentina, China," *Nucleonics Week*, September 24, 1992, pp. 2–3.
Ho, Daniel, Kosuke Imai, Gary King, and Elizabeth A. Stuart, "Matching as Nonparametric Preprocessing for Reducing Model Dependence in Parametric Causal Inference," *Political Analysis*, Vol. 15 (2007), pp. 199–236.
Holloway, David, "Other People's Nukes: Review of Spying on the Bomb," *New York Times*, March 26, 2006.
——, *Stalin and the Bomb: The Soviet Union and Atomic Energy, 1939–1956* (New Haven: Yale University Press, 1994).
Horne, Alistair, *Savage War of Peace: Algeria, 1954–1962* (New York: NYRB Classics, 2006).
Horowitz, Michael, "The Spread of Nuclear Weapons and International Conflict: Does Experience Matter?" *Journal of Conflict Resolution*, Vol. 53, No. 2 (April 2009).

———, "Who's behind that Curtain? Unveiling Potential Leverage over Pyongyang," *Washington Quarterly*, Vol. 28, No. 1 (Winter 2004/2005), pp. 21–44.
Huntington, Samuel P., *The Clash of Civilizations and the Remaking of World Order* (New York: Simon and Shuster, 1998).
Huth, Paul K., "Deterrence and International Conflict: Empirical Findings and Theoretical Debates," *Annual Review of Political Science*, Vol. 2 (1999), pp. 25–48.
———, *Extended Deterrence and the Prevention of War* (New Haven: Yale University Press, 1988).
———, "Reputations and Deterrence: A Theoretical and Empirical Assessment," *Security Studies*, Vol. 7, No. 1 (1997), pp. 72–99.
Hymans, Jacques E. C., *The Psychology of Nuclear Proliferation: Identity, Emotions, and Foreign Policy* (Cambridge: Cambridge University Press, 2006).
Institute for Defense and Disarmament Studies, *Nonproliferation Treaty*, available at www.idds.org/issNucTreatiesNPT.html.
Inter-American Treaty of Reciprocal Assistance, available at www.oas.org/juridico/english/Treaties/b-29.html.
Jabko, Nicholas, and Steven Weber, "A Certain Idea of Nuclear Weapons: France's Non-Proliferation Policies in Theoretical Perspective," *Security Studies*, Vol. 8, No. 1 (Fall 1998), pp. 108–150.
Jacobsen, John Kurt, and Claus Hofhansel, "Safeguards and Profits: Civilian Nuclear Exports, Neo-Marxism, and the Statist Approach," *International Studies Quarterly*, Vol. 28, No. 2 (June 1984), pp. 204–205.
Jaggers, K., and T. R. Gurr, "Tracking Democracy's Third Wave with the Polity III Data," *Journal of Peace Research*, Vol. 32, No. 4 (1995), pp. 469–482.
Jentleson, Bruce W., and Christopher A. Whytock, "Who Won Libya? The Force-Diplomacy Debate and its Implications for Theory and Policy," *International Security*, Vol. 30, No. 3 (Winter 2006), pp. 47–86.
Jervis, Robert, "Bargaining and Bargaining Tactics," in James Roland Pennock and John W. Chapman, eds., *Coercion* (Chicago: Aldine-Atherton, 1972).
———, "Deterrence Theory Revisited," *World Politics*, Vol. 31, No. 2 (January 1979), pp. 303–304.
———, *The Illogic of American Nuclear Strategy* (Ithaca, NY: Cornell University Press, 1985).
———, *The Meaning of the Nuclear Revolution: Statecraft and the Prospect of Armageddon* (Ithaca, NY: Cornell University Press, 1989).
———, *Perception and Misperception in International Politics* (Princeton, NJ: Princeton University Press, 1976).
———, "The Political Effects of Nuclear Weapons: A Comment," *International Security*, Vol. 13, No. 2 (Fall 1988), p. 80.
———, "Realism, Neoliberalism, and Cooperation: Understanding the Debate," *International Security*, Vol. 24, No. 1 (Summer 1999), pp. 42–63.
———, "Security Regimes," in Stephen Krasner, ed., *International Regimes* (Ithaca, NY: Cornell University Press, 1983).
Jervis, Robert, Richard Ned Lebow, and Janice Stein, *Psychology and Deterrence* (Baltimore: Johns Hopkins University, 1989).
Jo, Dong-Joon, and Erik Gartzke, "Determinants of Nuclear Weapons Proliferation: A Quantitative Model," *Journal of Conflict Resolution*, Vol. 51, No. 1 (February 2007), pp. 167–194.
Johnston, Alastair Iain, "International Structures and Chinese Foreign Policy," in Samuel S. Kim, ed., *China and the World: Chinese Foreign Policy Faces the New Millennium* (New York: Westview, 1998).

Jones, Rodney W., "Studying the Emerging Nuclear Suppliers," in William C. Potter, ed., *International Nuclear Trade and Nonproliferation* (Lexington, MA: Lexington Books, 1990).

Jones, Rodney W., and Mark G. McDonough with Toby F. Dalton and Gregory D. Koblentz, *Tracking Nuclear Proliferation: A Guide in Maps and Charts* (Washington, D.C.: Carnegie Endowment for International Peace, 1998).

Jones, Rodney W., Cesare Merlini, Joseph F. Pilat, and William C. Potter, eds., *The Nuclear Suppliers and Nonproliferation: International Policy Choices* (Lexington, MA: Lexington Books, 1985).

Kampani, Guarav, "Proliferation Unbound: Nuclear Tales from Pakistan," Center for Nonproliferation Studies Research Story, Monterey Institute of International Studies, February 23, 2004, available at http://cns.miis.edu/pubs/week/040223.htm.

Kaplan, Fred, "The Unspeakable Truth: What Bush Dares Not Say about North Korea," *Slate*, January 7, 2003.

Kaplan, Morton A., *System and Process in International Politics* (London: John Wiley and Sons, 1965).

Kapur, Ashok, *Pakistan's Nuclear Development* (New York: Croom Helm, 1987).

Kapur, S. Paul, *Dangerous Deterrent: Nuclear Weapons Proliferation and Conflict in South Asia* (Stanford, CA: Stanford University Press, 2007).

———, "India and Pakistan's Unstable Peace: Why Nuclear South Asia Is Not Like Cold War Europe," *International Security*, Vol. 30, No. 2 (Fall 2005), pp. 127–152.

Karamat, Jehangir, Pakistani Ambassador to the United States, interview with the author, Berkeley, California, March 2006.

Karl, David J., "Proliferation Pessimism and Emerging Nuclear Powers," *International Security*, Vol. 21, No. 3 (Winter 1996–1997), pp. 87–119.

Karnad, Bharat, *Nuclear Weapons and Indian Security: The Realist Foundations of Strategy* (New Delhi: Macmillan India Limited, 2002).

Katz, James Everett, and Onkar S. Marwah, eds., *Nuclear Power in Developing Countries* (Lexington, MA: Lexington Books, 1982).

Keck, Margaret, and Kathryn Sikkink, *Activists beyond Borders: Advocacy Networks in International Politics* (Ithaca, NY: Cornell University Press, 1998).

Kennedy, John F., Letter to Ben Gurion, in State Department Deptel 780 (Tel Aviv), May 4, 1963, NSF, Box 119a, John F. Kennedy Library.

Keohane, Robert O., *After Hegemony: Cooperation and Discord in the World Political Economy* (Princeton, NJ: Princeton University Press, 1984).

Kessler, Glenn, "North Korea May Have Sent Libya Nuclear Material, U.S. Tells allies," *Washington Post*, February 2, 2005, p. A01.

Kessler, Glenn, and Robin Wright, "Edwards Says Kerry Plans to Confront Iran on Weapons," *Washington Post*, August 30, 2004, p. A1.

"Khan: Musharraf in on North Korea Nuke Deal," *Associated Press*, July 4, 2008.

Khan, Zeba, "Abdul Kadeer Khan: The Man Behind the Myth," Interview, available at www.yespakistan.com/people/abdul_qadeer.asp.

Khong, Yuen Foong, *Analogies at War* (Princeton, NJ: Princeton University Press, 1992).

Kibaroglu, Mustafa, "Good for the Shah, Banned for the Mullahs," *Middle East Journal*, Vol. 60, No. 2 (Spring 2006).

King, Gary, Robert O. Keohane, and Sidney Verba, *Designing Social Inquiry: Scientific Inference in Qualitative Research* (Princeton, NJ: Princeton University Press, 1994).

Kier, Elizabeth, "Culture and Military Doctrine: France between the Wars," *International Security*, Vol. 19, No. 4 (Spring 1995), pp. 65–93.

Kier, Elizabeth and Jonathan Mercer, "Setting Precedents in Anarchy: Military Intervention and Weapons of Mass Destruction," *International Security*, Vol. 20, No. 4 (Spring 1996), pp. 77–106.

King, Gary, Michael Tomz, and Jason Wittenberg, "Making the Most of Statistical Analyses: Improving Interpretation and Presentation," *American Journal of Political Science*, Vol. 44, No. 2 (April 2000), pp. 341–355.

King, Gary, and Langche Zeng, "Logistic Regression in Rare Events Data," *Political Analysis*, Vol. 9, No. 2 (Spring 2001), pp. 137–163.

Kinsella, David, "Conflict in Context: Superpower Arms Transfers and Third World Rivalry during the Cold War," *American Journal of Political Science*, Vol. 38, No. 3 (August 1994), pp. 557–581.

Kinsella, David, and Herbert K. Tillema, "Arms and Aggression in the Middle East: Overt Military Interventions, 1948–1991," *Journal of Conflict Resolution*, Vol. 39, No. 2 (June 1995), pp. 306–329.

Kissinger, Henry A., *The White House Years* (Boston: Little, Brown, 1979).

——, *Years of Upheaval* (Boston: Little, Brown, 1982).

Klein, James P., Gary Goertz, and Paul F. Diehl, "The New Rivalry Dataset: Procedures and Patterns," *Journal of Peace Research*, Vol. 43, No. 3 (May 2006), pp. 331–348.

Knopf, Jeffrey W., "Recasting the Optimism/Pessimism Debate," *Security Studies*, Vol. 12, No. 1 (Autumn 2002), pp. 41–96.

Knorr, Klaus, "Limited Strategic War," in Klaus Knorr and Thorton Read, eds., *Limited Strategic War* (Princeton, NJ: Princeton University Press, 1962).

Knorr, Klaus, and Thorton Read, eds., *Limited Strategic War* (Princeton, NJ: Princeton University Press, 1962).

Koch, Andrew, "Khanfessions of a Proliferator," *Jane's Defense Weekly*, March 3, 2004.

——, "Yugoslavia's Nuclear Legacy: Should We Worry?" *Nonproliferation Review*, Vol. 4, No. 3 (Spring/Summer 1997), pp. 123–128.

Koch, Andrew, and Jenette Wolf, "Iran's Nuclear Procurement Program," *Nonproliferation Review*, Vol. 5, No. 1 (Fall 1997), pp. 123–135.

Kohl, Wilfred L., *French Nuclear Diplomacy* (Princeton, NJ: Princeton University Press, 1971).

Koremenos, Barbara, Charles Lipson, and Duncan Snidal, "The Rational Design of International Institutions," *International Organization*, Vol. 55, No. 4 (Autumn 2001), pp. 761–799.

Krasner, Stephen, ed., *International Regimes* (Ithaca, NY: Cornell University Press, 1983).

——, "State Power and the Structure of International Trade," *World Politics*, Vol. 28, No. 3 (April 1976), pp. 317–347.

Krasno, Jean, "Brazil's Secret Nuclear Program," *Orbis*, Vol. 38, No. 3 (Summer 1994), pp. 425–436.

Kreutzmann, Hermann, "The Karakoram Highway: The Impact of Road Construction on Mountain Societies," *Modern Asian Studies*, Vol. 25, No. 4 (October 1991), pp. 711–736.

Kroenig, Matthew, *The Enemy of My Enemy Is My Customer: Why States Provide Sensitive Nuclear Assistance*, Ph.D. dissertation, University of California, Berkeley, 2007.

——, "Exporting the Bomb: Why States Provide Sensitive Nuclear Assistance," *American Political Science Review*, Vol. 103, No. 1 (February 2009).

——, "Importing the Bomb: Sensitive Nuclear Assistance and Nuclear Proliferation," *Journal of Conflict Resolution*, Vol. 53, No. 2 (April 2009).

Kroenig, Matthew, and Jay Stowsky, "War Makes the State, But Not As It Pleases: Homeland Security and American Anti-Statism," *Security Studies*, Vol. 15, No. 2 (April–June 2006), pp. 225–270.

Lamb, Alastair, *The Sino-Indian Border in Ladakh* (Canberra: Australian National University Press, 1973).

Langewiesche, William, *The Atomic Bazaar: The Rise of the Nuclear Poor* (New York: Farrar Straus and Giroux, 2007).

——, "The Point of No Return," *Atlantic Monthly,* Vol. 297, No. 1 (January/February 2006), pp. 96–97.
——, "The Wrath of Khan," *Atlantic Monthly,* Vol. 296, No. 4 (November 2005), pp. 62–85.
Laufer, Robert, "Argentina Looks to Reprocessing to Fill its Own Needs Plus Plutonium Sales," *Nuclear Fuel,* November 8, 1982, p. 3.
Lavoy, Peter R., "The Enduring Effects of Atoms for Peace," *Arms Control Today* (December 2003).
——, "Nuclear Myths and the Causes of Nuclear Proliferation," in Zachary S. Davis and Benjamin Frankel, eds., *The Proliferation Puzzle: Why Nuclear Weapons Spread (and What Results)* (Portland: Frank Cass, 1993).
——, "The Strategic Consequences of Nuclear Proliferation," *Security Studies,* Vol. 4, No. 4 (Summer 1995), pp. 695–753.
Lavoy, Peter R., Scott D. Sagan, and James J. Wirtz, eds., *Planning the Unthinkable: How New Powers Will Use Nuclear, Chemical, and Biological Weapons* (Ithaca, NY: Cornell University Press, 2000).
Lebovic, James, *Deterring International Terrorism and Rogue States: U.S. National Security Policy after 9/11* (New York: Routledge, 2006).
Legro, Jeffrey, "Which Norms Matter? Revisiting the 'Failure' of Institutionalism," *International Organization,* Vol. 51, No. 1 (Winter 1997), pp. 31–63.
Lesch, David, ed., *The Middle East and the United States: A Historical and Political Reassessment* (New York: Westview Press, 2006).
Lester, Richard K., "U.S.-Japanese Nuclear Relations: Structural Change and Political Strain," *Asian Survey,* Vol. 22, No. 5 (May 1982), pp. 417–433.
Levi, Michael, "Deterring Nuclear Terrorism," *Issues in Science and Technology,* Vol. 20, No. 3 (Spring 2004), available at www.issues.org/20.3/levi.html.
——, *On Nuclear Terrorism* (Cambridge, MA: Harvard University Press, 2007).
Levine, Paul, and Ron Smith, "Arms Export Controls and Proliferation," *Journal of Conflict Resolution,* Vol. 44, No. 6 (December 2000), pp. 885–895.
Levite, Ariel, former official, Israeli Atomic Energy Commission, interview with the author, Stanford, California, June 2007.
Levite, Ariel, and Emily B. Landau, "Arab Perceptions of Israel's Nuclear Posture, 1960–1967," *Israel Studies,* Vol. 1, No. 1 (Spring 1996), pp. 34–59.
Levy, Adrian, and Katherine Scott Clark, *Deception: Pakistan, the United States, and the Secret Trade in Nuclear Weapons* (New York: Walker and Company, 2007).
Lewis, John W., and Xue Litai, *China Builds the Bomb* (Stanford, CA: Stanford University Press, 1988).
Liberman, Peter, "The Rise and Fall of the South African Bomb," *International Security,* Vol. 26, No. 2 (Summer 2001), pp. 45–86.
——, "Israel and the South African Bomb," *Nonproliferation Review,* Vol. 11, No. 2 (Summer 2004), pp. 46–80.
Linzer, Dafna, "U.S. Misled Allies about Nuclear Export: North Korea Sent Material to Pakistan, Not Libya," *New York Times,* March 20, 2005, A01.
Linzer, Dafna, and Molly Moore, "U.N. Agency Finds Iran Noncompliant," *Washington Post,* April, 28, 2006, p. A16.
Little, Douglas, "The Making of a Special Relationship: The United States and Israel, 1957–68," *International Journal of Middle East Studies,* No. 25 (1993), pp. 563–585.
Logevall, Fredrik, and Andrew Preston, eds., *Nixon in the World: American Foreign Relations, 1969–1977* (New York: Oxford University Press, 2008).
Lonroth, Mans, and William Walker, *Nuclear Power Struggles: Industrial Competition and Proliferation Control* (London: George, Allen, and Unwin, 1983).

———, *The Viability of the Civilian Nuclear Industry* (New York and London: International Consultative Group on Nuclear Energy, jointly sponsored by the Rockefeller Foundation and the Royal Institute on International Affairs, 1979).
Louis, William Roger, and Roger Owen, eds., *Suez 1956: The Crisis and Its Consequences* (Oxford: Clarendon, 1989).
Love, Robert W., Jr., *History of the U.S. Navy (1942–1991)* (Harrisburg, PA: Stackpole Books, 1992).
Machiavelli, Niccolò, *The Prince,* translated and edited by Robert M. Adams (New York: W.W. Norton & Company, Inc., 1977).
MacKenzie, Donald, and Graham Spinardi, "Tacit Knowledge, Weapons Design, and the Uninvention of Nuclear Weapons," *American Journal of Sociology,* Vol. 100 (1995), pp. 44–99.
Mansfield, Edward D., and Jack Snyder, "Democratization and the Danger of War," *International Security,* Vol. 20, No. 1 (Summer 1995), pp. 5–38.
Martin, André, "The Military and Political Contradictions of the Suez Affair: A French Perspective," in Selwyn Ilan Troen and Moshe Shemesh, eds., *The Suez Crisis 1956: Retrospective and Reappraisal* (New York: Columbia University Press, 1990).
Martin, Michel L., *Warriors to Managers* (Chapel Hill: University of North Carolina Press, 1981).
Maxwell, Neville, *India's China War* (New York: Pantheon Books, 1970).
McConnell, Michael J., "Annual Threat Assessment of the Director of National Intelligence for the Senate Armed Services Committee," February 27, 2007, unclassified Statement for the Record.
Mearsheimer, John J., "Back to the Future: Instability in Europe after the Cold War," *International Security,* Vol. 15, No. 1 (Summer 1990), pp. 5–56.
———, "The Case for a Ukrainian Nuclear Deterrent," *Foreign Affairs,* Vol. 72, No. 3 (Summer 1993), pp. 50–66.
———, "The False Promise of International Institutions," *International Security,* Vol. 19, No. 3 (Winter 1994/1995), pp. 5–49.
———, *The Tragedy of Great Power Politics* (New York: W.W. Norton, 2001).
Mearsheimer, John J., and Stephen Walt, *The Israel Lobby and U.S. Foreign Policy* (New York: Farrar, Strauss, and Giroux, 2007).
———, "An Unnecessary War," *Foreign Policy,* No. 134 (January/February 2003), pp. 50–59.
Medeiros, Evan S., *Chasing the Dragon: Assessing China's System of Export Controls for WMD-Related Goods and Technologies* (Santa Monica, CA: RAND, 2005).
Mercer, Jonathan, *Reputation and International Politics* (Ithaca, NY: Cornell University Press, 1996).
Meyer, Stephen M., *The Dynamics of Nuclear Proliferation* (Chicago: University of Chicago Press, 1984).
Miller, Steve E., "Assistance to Newly Proliferating Nations," in Robert D. Blackwill and Albert Carnesale, eds., *New Nuclear Nations: Consequences for U.S. Policy* (New York: Council on Foreign Relations Press, 1994), pp. 97–131.
———, "The Case against a Ukrainian Nuclear Deterrent," *Foreign Affairs,* Vol. 73, No. 3 (Summer 1993), pp. 67–80.
Moaz, Zeev, and Bruce Russett, "Normative and Structural Causes of Democratic Peace, 1946–1987," *American Political Science Review,* Vol. 87, No. 3 (September 1993), pp. 624–638.
Mohan, C. Raja, *Crossing the Rubicon: The Shaping of India's Foreign Policy* (New York: Palgrave Macmillan, 2004).
———, *Impossible Allies: Nuclear India, the United States, and Global Order* (New Delhi: India Research Press, 2006).

Montgomery, Alexander H., "Ringing in Proliferation: How to Dismantle an Atomic Bomb Network," *International Security,* Vol. 30, No. 2 (Fall 2005), pp. 153–187.
Montgomery, Alexander H., and Scott D. Sagan, "The Perils of Predicting Proliferation," *Journal of Conflict Resolution,* Vol. 53, No. 2 (April 2009).
Moravcsik, Andrew, "Armaments among Allies: European Weapons Collaboration, 1975–1985," in Peter Evans, Harold Jacobson, and Robert Putnam, eds., *Double-Edge Diplomacy: International Bargaining and Domestic Politics* (Berkeley: University of California Press, 1993).
Morgenthau, Hans J., *Politics among Nations* (New York: Knopf, 1948).
Morris, Harvey, "Hamas Rocket Squads Resume Israeli Attacks," *Financial Times,* June 11, 2006, available at http://news.ft.com/cms/s/3d854060-f96c-11da-8ced-0000779e2340.html.
Morstein, Jennifer Hunt, and Wayne D. Perry, "Commercial Nuclear Trading Networks as Indicators of Nuclear Weapons Intentions," *Nonproliferation Review,* Vol. 7, No. 3 (Fall/Winter 2000), pp. 75–91.
Most, Benjamin A., and Harvey Starr, *Inquiry, Logic, and International Politics* (Columbia: University of South Carolina Press, 1989).
Milhollin, Gary, "The Iraqi Bomb," *New Yorker,* February 1, 1993, p. 47.
Milstein, Uri, *History of the War of Independence,* Vols. 1–4 (New York: University Press of America, 1996).
Mueller, John, "The Essential Irrelevance of Nuclear Weapons," *International Security,* Vol. 13, No. 2 (Fall 1988), pp. 3–17.
Mullins, Martin, *In the Shadow of Generals: Foreign Policy Making in Argentina, Brazil, and Chile* (Burlington, VT: Ashgate, 2006).
Musharraf, Pervez, *In the Line of Fire: A Memoir* (New York: Free Press, 2006).
"Musharraf Lied about A.Q.'s Role in N-Program Says Hendrina," *Daily Times* (Pakistan), August 10, 2008.
Myers, Ramon H., and Jialin Zhang, *The Struggle across the Taiwan Strait: The Divided China Problem* (Stanford, CA: Hoover Institution Press, 2006).
National Defense University, "U.S. Military Bases in the Cold War," available at www.ndu.edu.
National Intelligence Council, 2020 Project, "Mapping the Global Future," December 2004.
"National Security Strategy of the United States of America," March 2006, available at www.whitehouse.gov/nsc/nss/2006/sectionV.html.
Nalebuff, Barry, "Brinkmanship and Nuclear Deterrence: The Neutrality of Escalation," *Conflict Management and Peace Science,* Vol. 9, No. 2 (Spring 1986), pp. 19–30.
Nayar, Baldav Raj, and T. V. Paul, *India in the World Order: Searching for Major Power Status* (Cambridge, U.K.: Cambridge University Press, 2002).
Neustadt, Richard E., and Ernest R. May, *Thinking in Time: The Uses of History for Decision Makers* (New York: Free Press, 1986).
Nichol, J. P., and G. L. McDaniel, "Yugoslavia," in James Everett Katz and Onkar S. Marwah, eds., *Nuclear Power in Developing Countries* (Lexington, MA: Lexington Books, 1982).
Nogee, Joseph L., *Soviet Nuclear Proliferation Policy* (Carlisle, PA: U.S. Army War College, 1980).
Norris, Robert S., Hans M. Kristensen, and Joshua Handler, "Pakistan's Nuclear Forces, 2001," *Bulletin of the Atomic Scientists,* Vol. 58, No. 1 (January/February 2002), pp. 70–71.
———, "North Korea's Nuclear Program, 2003," *Bulletin of the Atomic Scientists,* Vol. 59, No. 2 (March/April 2003), pp. 74–77.

Nuclear Energy Agency, *Decommissioning in NEA Member Countries, Current Status: Norway,* available at www.nea.fr/html/rwm/wpdd/norway.pdf.
——, *Decommissioning of Nuclear Installations in Italy* (January 2006), available at www.nea.fr/html/rwm/wpdd/italy.pdf.
Nuclear Proliferation Prevention Act of 1994, Title VIII, Sec. 825, Foreign Relations Authorization Act, Fiscal Year 1994 and 1995, P.L. 103–236 (1994).
Nuclear Suppliers Group Guidelines for Nuclear Transfers, available at www.fas.org/nuke/control/nsg/text/inf254.htm.
Nuclear Threat Initiative, *Country Profiles,* available at http://nti.org/e_research/profiles/index.html.
——, *Russia Nuclear Exports to Iran: Reactors,* available at www.nti.org/db/nisprofs/russia/exports/rusiran/react.htm.
Office of the Director of National Intelligence, *Iran: Nuclear Intentions and Capabilities,* November 2007, available at www.dni.gov/press_releases/20071203_release.pdf.
Office of the Secretary of Defense, *2008 Annual Report on the Military Power of the People's Republic of China,* available at www.defenselink.mil/pubs/pdfs/China_Military_Report_08.pdf.
——, *The National Defense Strategy of the United States of America,* March 2005, available at www.defenselink.mil/news/Mar2005/d20050318nds1.pdf.
Ogawa, Shinichi, and Michael Schiffer, "Japan's Plutonium Reprocessing Dilemma," *Arms Control Today* (October 2005), available at www.armscontrol.org/act/2005_10/Oct-Japan.asp.
Ogilvie-White, Tanya, "Is There A Theory of Nuclear Proliferation? An Analysis of the Contemporary Debate," *Nonproliferation Review,* Vol. 4, No. 1 (Fall 1996), pp. 43–60.
Oneal, John R., and Bruce Russett, "The Classical Liberals Were Right: Democracy, Interdependence, and Conflict, 1950–1985," *International Studies Quarterly,* Vol. 41, No. 2 (June 1997), pp. 267–294.
O'Neill, Barry, *Honor, Symbols, and War* (Ann Arbor: University of Michigan Press, 1999).
Orlov, Vladimir A., and Alexander Vinnikov, "The Great Guessing Game: Russia and the Iranian Nuclear Issue," *Washington Quarterly,* Vol. 28, No. 2 (Spring 2005), pp. 49–66.
Pape, Robert A., *Bombing to Win: Air Power and Coercion in War* (Ithaca, NY: Cornell University Press, 1996).
——, "Soft Balancing against the United States," *International Security,* Vol. 30, No. 5 (Fall 2005), pp. 7–45.
Paribatra, Sukhumbhand, *The Taiwan Straits Crisis of 1958: A Study of the Use of Naval Power* (Bangkok: Institute of Asian Studies, Chulalongkorn University, 1981).
Parsi, Trita, *Treacherous Alliance: The Secret Dealings of Israel, Iran, and the United States* (New Haven: Yale University Press, 2007).
Paul, T. V., "Chinese-Pakistani Nuclear/Missile Ties and Balance of Power Politics," *Nonproliferation Review,* Vol. 10, No. 2 (Summer 2003), pp. 21–29.
—— ed., *The India-Pakistan Conflict: An Enduring Rivalry* (Cambridge, U.K.: Cambridge University Press, 2005).
——, *Power Versus Prudence: Why Nations Forgo Nuclear Weapons* (Montreal: McGill-Queen's University Press, 2000).
——, "Soft Balancing in the Age of U.S. Primacy," *International Security,* Vol. 30, No. 5 (Fall 2005), pp. 46–71.
——, "Systemic Conditions and Security Cooperation: Explaining the Persistence of the Nuclear Non-Proliferation Regime," *Cambridge Review of International Affairs,* Vol. 16, No. 1 (April 2003), pp. 135–155.
Péan, Pierre, *Les Deux Bombes* (Paris: Fayard, 1981).

Perkovich, George, *India's Nuclear Bomb: The Impact on Global Proliferation* (Berkeley: University of California Press, 1999).
Perrow, Charles, *Normal Accidents: Living with High Risk Technologies* (New York: Basic Books, 1984).
Peres, Shimon, *Battling for Peace: A Memoir* (London: Weidenfeld and Nicolson, 1995).
———, *David's Sling* (London: Weidenfeld and Nicolson, 1970).
———, *From These Men: Seven Founders of the State of Israel* (New York: Wyndham, 1979).
———, "The Road to Sèvres," in Selwyn Ilan Troen and Moshe Shemesh, eds., *The Suez-Sinai Crisis 1956: Retrospective and Reappraisal* (New York: Columbia University Press, 1990).
Pierre, Andrew J., *The Global Politics of Arms Sales* (Princeton, NJ: Princeton University Press, 1982).
———, *Nuclear Politics: The British Experience with an Independent Strategic Force, 1939–1970* (London: Oxford University Press, 1972).
Pierson, Paul, "Path Dependence, Increasing Returns, and the Study of Politics," *American Political Science Review*, Vol. 94, No. 2 (June 2000), pp. 251–267.
Pinkus, Benjamin, "Atomic Power to Israel's Rescue: French-Israeli Nuclear Cooperation, 1949–1957," *Israel Studies*, Vol. 7, No. 1 (2002), pp. 104–138.
Pilat, Joseph F., "The French, Germans, and Japanese and the Future of the Nuclear Supply Regime," in Rodney W. Jones, Cesare Merlini, Joseph F. Pilat, and William C. Potter, eds., *The Nuclear Suppliers and Nonproliferation: International Policy Choices* (Lexington, MA: Lexington Books, 1985), pp. 81–92.
Podvig, Pavel, eds., *Russian Strategic Nuclear Forces* (Cambridge, MA: MIT Press, 2001).
Posen, Barry R., "Command of the Commons: The Military Foundation of U.S. Hegemony," *International Security*, Vol. 28, No. 1 (Summer 2003), pp. 5–46.
———, *Inadvertent Escalation: Conventional War and Nuclear Risks* (Ithaca, NY: Cornell University Press, 1991).
———, "The Security Dilemma and Ethnic Conflict," *Survival*, Vol. 35, No. 1 (Spring 1993), pp. 27–47.
———, "U.S. Nuclear Policy in a Nuclear-Armed World, or What If Iraq Had Had Nuclear Weapons?" in Victor Utgoff, ed., *The Coming Crisis* (Boston: MIT Press, 2000), pp. 57–190.
———, "U.S. Security Policy in a Nuclear-Armed World (Or: What If Iraq Had Had Nuclear Weapons?)," *Security Studies*, Vol. 6, No. 3 (1997), pp. 1–31.
———, "We Can Live with a Nuclear Iran," *New York Times*, February 27, 2006.
Potter, William C., ed., *International Nuclear Trade and Nonproliferation: The Challenge of the Emerging Nuclear Suppliers* (Lexington, MA: Lexington Books, 1990).
———, *Nuclear Power and Nonproliferation: An Interdisciplinary Perspective* (Cambridge, MA: Oelgeschlager, Gunn, and Hain, 1982).
———, "U.S.-Soviet Cooperative Measures for Nonproliferation," in Rodney W. Jones, Cesare Merlini, Joseph F. Pilat, and William C. Potter, eds., *The Nuclear Suppliers and Nonproliferation: International Policy Choices* (Lexington, MA: Lexington Books, 1985).
Potter, William C., Djuro Miljanic, and Ivo Slaus, "Tito's Nuclear Legacy," *Bulletin of Atomic Scientists*, Vol. 56, No. 2 (March/April 2000), pp. 63–70.
Powaski, Ronald E., *The Cold War: The United States and the Soviet Union, 1917–1991* (New York: Oxford University Press, 1997).
Powell, Robert, "Absolute and Relative Gains in International Relations Theory," *American Political Science Review*, Vol. 85, No. 4 (December 1991), pp. 1303–1320.
———, "Anarchy in International Relations Theory: The Neo-Realist-Neo-Liberal Debate," *International Organization*, Vol. 48, No. 2 (Spring 1994), pp. 313–344.

———, "Crisis Bargaining, Escalation, and MAD," *American Political Science Review*, Vol. 81, No. 3 (September 1987), pp. 717–736.
———, "Guns, Butter, and Anarchy," *American Political Science Review*, Vol. 87, No. 1 (March 1993), pp. 115–132.
———, "Nuclear Deterrence Theory, Nuclear Proliferation, and National Missile Defense," *International Security*, Vol. 27, No. 4 (Spring 2003), pp. 86–118.
———, *Nuclear Deterrence Theory: The Search for Credibility* (Cambridge, U.K.: Cambridge University Press, 1990).
———, "The Theoretical Foundations of Nuclear Deterrence," *Political Science Quarterly*, Vol. 100, No. 1 (Spring 1985), pp. 75–96.
"President Bush Outlines Iraqi Threat," October 7, 2002, available at www.whitehouse.gov.
"President Bush's Statement on North Korea Nuclear Test," October 9, 2006, available at www.whitehouse.gov/news/releases/2006/10/20061009.html.
"The President's State of the Union Address," January 29, 2002, available at www.whitehouse.gov/news/releases/2002/01/20020129-11.html.
Press, Daryl, "The Credibility of Power: Assessing Threats during the 'Appeasement' Crises of the 1930s," *International Security*, Vol. 29, No. 2 (Winter 2004/05), pp. 136–169.
———, *Calculating Credibility: How Leaders Assess Military Threats* (Ithaca, NY: Cornell University Press, 2005).
Pritchard, Charles L., *Failed Diplomacy: The Tragic Story of How North Korea Got the Bomb* (Washington, D.C.: Brookings Institution Press, 2007).
Przeworski, Adam, and Henry Teune, *Logic of Comparative Social Inquiry* (New York: Krieger, 1982).
Puchala, Donald J., and Raymold E. Hopkins, "International Regimes: Lessons from Inductive Analysis," in Stephen Krasner, ed., *International Regimes* (Ithaca, NY: Cornell University Press, 1983).
Quester, George, *The Politics of Nuclear Proliferation* (Baltimore: Johns Hopkins University Press, 1973).
———, "The Statistical 'N' of the 'Nth' Nuclear-Weapon States," *Journal of Conflict Resolution*, Vol. 27, No. 1 (March 1983), pp. 161–179.
Quinn, Frederick, *The French Overseas Empire* (New York: Praeger, 2001).
Rabinovich, Abraham, *The Yom Kippur War* (New York: Schocken Books, 2004).
Rauchhaus, Robert, "Evaluating the Nuclear Peace Hypothesis: A Quantitative Approach," *Journal of Conflict Resolution*, Vol. 53, No. 2 (April 2009).
Reed, Thomas C., and Danny B. Stillman, *The Nuclear Express: A Political History of the Bomb and Its Proliferation* (Minneapolis: Zenith Press, 2009).
Reiss, Mitchell, *Bridled Ambition: Why Countries Constrain their Nuclear Capabilities* (Princeton, NJ: Princeton University Press, 1995).
———, *Without the Bomb: The Politics of Nuclear Nonproliferation* (New York: Columbia University Press, 1988).
Reiter, Dan, and Allan C. Stam, *Democracies at War* (Princeton, NJ: Princeton University Press, 2002).
———, "Democracy, War Initiation, and Victory," *American Political Science Review*, Vol. 92, No. 2 (June 1998), pp. 377–389.
Rhodes, Richard, *The Making of the Atomic Bomb* (New York: Simon and Schuster, 1995).
Richelson, Jeffrey T., *Spying on the Bomb: American Nuclear Intelligence from Nazi Germany to Iran and North Korea* (New York: W.W. Norton, 2006).
Richter, Paul, and Greg Miller, "U.S. Offers Evidence of North Korea-Syria Nuclear Plant," *Los Angeles Times*, April 25, 2008.

Rigby, Randall L., *The Soviet Union and Nuclear Proliferation* (Carlisle, PA: U.S. Army War College, 1988).
Robertson, Terrence, *Crisis: The Inside Story of the Suez Conspiracy* (Toronto: McClelland and Stewart, 1964).
"Rocket Man v. Bulldozer," *Economist,* April 5–11, 2008, p. 45.
Rodrik, Dani, *Has Globalization Gone Too Far?* (Washington, D.C.: Institute for International Economics, 1997).
Rogers, Barbara, and Zedenk, Cervenka, *The Nuclear Axis: Secret Collaboration between West Germany and South Africa* (New York: Times Books, 1978).
Ruina, Jack, Louis Alvarez, William Donn, Richard Garwin, Riccardo Giacconi, Richard Muller, Wolfgang Panofsky, Allen Peterson, and Williams Sarles, "Ad hoc Panel Report on the September 22 Event," Executive Office of the President, Office of Science and Technology Policy, July 17, 1980, available at www.fas.org/rlg/800717-vela.pdf.
Russett, Bruce, *Grasping the Democratic Peace* (Princeton, NJ: Princeton University Press, 1993).
Rynning, Sten, *Changing Military Doctrine: Presidents and Military Power in Fifth Republic France, 1958–2000* (London: Praeger, 2002).
Sagan, Scott D., "The Commitment Trap: Why the United States Should Not Use Nuclear Threats to Deter Biological and Chemical Weapon Attacks," *International Security,* Vol. 24, No. 4 (Spring 2000), pp. 85–115.
———, "Dissuasion and the NPT Regime: Complementary or Contradictory Strategies?" *Strategic Insights,* Vol. 3, No. 10 (October 2004).
———, "Keeping the Bomb Away from Tehran," *Foreign Affairs,* Vol. 85, No. 5 (September/October 2006), pp. 45–59.
———, *The Limits of Safety: Organizations, Accidents, and Nuclear Weapons* (Princeton, NJ: Princeton University Press, 1993).
———, "More Will Be Worse," in Scott D. Sagan and Kenneth N. Waltz, eds., *The Spread of Nuclear Weapons: A Debate* (New York: W.W. Norton, 1995).
———, *Moving Targets: Nuclear Strategy and National Security* (Princeton, NJ: Princeton University Press, 1989).
———, "Rethinking the Causes of Nuclear Proliferation: Three Models in Search of a Bomb?" in Victor A. Utgoff, ed., *The Coming Crisis: Nuclear Proliferation, U.S. Interests, and World Order* (Cambridge, MA: MIT Press, 2000) pp. 17–50.
———, "Why Do States Build Nuclear Weapons? Three Models in Search of a Bomb," *International Security,* Vol. 21, No. 3 (Winter 1996/97), pp. 54–86.
Sagan, Scott D., and Kenneth Waltz, *The Spread of Nuclear Weapons: A Debate* (New York: W.W. Norton, 1995).
Sands, Amy, "Emerging Nuclear Suppliers: What's the Beef?" in William C. Potter, ed., *International Nuclear Trade and Nonproliferation* (Lexington, MA: Lexington Books, 1990).
Sanger, David, "Tested Early by North Korea, Obama Has Few Options," *New York Times,* May 29, 2009.
Sanjin, Gregory, "Promoting Stability or Instability? Arms Transfers and Regional Rivalries, 1950–1991," *International Studies Quarterly,* Vol. 43, No. 4 (December 1999), pp. 641–670.
Schafer, Ramsey, *The Statistical Sleuth* (Pacific Grove, CA: Duxbury, 2002).
Scheinman, Lawrence, *Atomic Energy Policy in France under the Fourth Republic* (Princeton, NJ: Princeton University Press, 1965).
———, "The Nuclear Fuel Cycle: A Challenge for Nonproliferation," *Disarmament Diplomacy,* No. 76 (March/April 2004).

Schelling, Thomas, *Arms and Influence* (New Haven: Yale University Press, 1966).
——, *The Strategy of Conflict* (Cambridge: Harvard University Press, 1960).
Schelling, Thomas, and Morton Halperin, *Strategy and Arms Control* (New York: Twentieth Century Fund, 1961).
Schiff, Benjamin N., *International Nuclear Technology Transfer: Dilemmas of Dissemination and Control* (London: Croom Helm, 1983).
Schweller, Randall L., "Bandwagoning For Profit: Bringing the Revisionist State Back In." *International Security*, Vol. 19, No. 1 (Summer 1994), pp. 72–107.
——, *Unanswered Threats: Political Constraints on the Balance of Power* (Princeton, NJ: Princeton University Press, 2006).
Seaborg, Glenn T., with Benjamin S. Loeb, *Stemming the Tide: Arms Control in the Johnson Years* (Lexington, MA: Lexington Books, 1987).
Sechser, Todd S., *Winning without a Fight: Power, Reputation, and Compellent Threats in International Crises*, Ph.D. dissertation, Stanford University, 2007.
Sekhon, Jasjeet S., "Matching: Algorithms and Software for Multivariate and Propensity Score Matching with Balance Optimization via Genetic Search," (2006), available at http://sekhon.berkeley.edu/matching.
——, "Multivariate and Propensity Score Matching Software with Automated Balance Optimization: The Matching Package for R," *Journal of Statistical Software*, forthcoming.
Sekhon, Jasjeet S., and Walter Mebane, "Genetic Optimization using Derivatives," *Political Analysis*, Vol. 7 (1998), pp. 187–210.
Seng, Jordan, "Less is More: Command and Control Advantages of Minor Nuclear States," *Security Studies*, Vol. 6, No. 4 (Summer 1997), pp. 50–92.
Share, Michael, *Where Empires Collide: Russian and Soviet Relations with Hong Kong, Taiwan, and Macao* (Hong Kong: Chinese University Press, 2007).
Sheffer, Gabriel, *Moshe Sharret: Biography of a Political Moderate* (Oxford: Oxford University Press, 1996), pp. 808–859.
Shuey, Robert, and Shirley A. Kan, "Chinese Missile and Nuclear Proliferation: Issues for Congress," *CRS Issue Brief 29*, September 9, 1995.
Siddiqa, Ayesha, *Military Inc.: Inside Pakistan's Military Economy* (New York: Pluto Press, 2007).
Siddiqa-Aqha, Ayesha, *Pakistan's Arms Procurement and Military Build-Up, 1979–1999: In Search of a Policy* (New York: Palgrave Macmillan, 2001).
Sikkink, Kathryn, "The Power of Principled Ideas: Human Rights Policies in the United States and Western Europe," in Judith Goldstein and Robert O. Keohane, eds., *Ideas and Foreign Policy: Beliefs, Institutions, and Political Change* (Ithaca, NY: Cornell University Press, 1993).
Simes, Dimitri K., "Soviet Policy toward the United States," in Joseph S. Nye Jr., ed., *The Making of America's Soviet Policy* (New Haven: Yale University Press, 1984).
Simmons, Beth A., and Daniel J. Hopkins, "The Constraining Power of International Treaties: Theory and Methods," *American Political Science Review*, Vol. 99, No. 4 (November 2005), pp. 623–633.
Singer, J. David, Stuart Bremer, and John Stuckey, "Capability, Distribution, Uncertainty, and Major Power War, 1820–1965," in Bruce Russett, ed., *Peace, War, and Numbers* (Beverly Hills, CA: Sage, 1972).
Singh, Sonali, and Christopher R. Way, "The Correlates of Nuclear Proliferation: A Quantitative Test," *Journal of Conflict Resolution*, Vol. 48, No. 6 (December 2004), pp. 859–885.
Sislin, John, "Arms as Influence: The Determinants of Successful Influence," *Journal of Conflict Resolution*, Vol. 38, No. 4 (December 1994), pp. 655–689.

Smith, Holland M., "The Development of Amphibious Tactics in the U.S. Navy," *Occasional Paper* (Washington, D.C.: History and Museums Division, U.S. Marine Corps, 1992).
Snidal, Duncan, "The Limits of Hegemonic Stability Theory," *International Organization*, Vol. 39, No. 4 (Autumn 1985), pp. 579–614.
———, "Relative Gains and the Pattern of International Cooperation," *American Political Science Review*, Vol. 85, No. 3 (September 1991), pp. 701–726.
Snyder, Glenn H., "The Balance of Power and the Balance of Terror," in Paul Seabury, ed., *The Balance of Power* (San Francisco: Chandler, 1965).
———, "The Security Dilemma in Alliance Politics," *World Politics*, Vol. 36, No. 4 (July 1984), pp. 461–495.
Snyder, Glenn, and Paul Diesing, *Conflict among Nations: Bargaining, Decision Making, and System Structure in International Crises* (Princeton, NJ: Princeton University Press, 1978).
Snyder, Jack, *Myths of Empire: Domestic Politics and International Ambition* (Ithaca, NY: Cornell University Press, 1993).
Solingen, Etel, "The Political Economy of Nuclear Restraint," *International Security*, Vol. 19, No. 2 (Fall 1994), pp. 126–169.
———, *Nuclear Logics: Contrasting Paths in East Asia and the Middle East* (Princeton, NJ: Princeton University Press, 2007).
———, *Regional Orders at Century's Dawn: Global and Domestic Influences on Grand Strategy* (Princeton, NJ: Princeton University Press, 1998).
Soman, Appu K., *Double-Edged Sword: Nuclear Diplomacy in Unequal Conflicts, the United States and China, 1950–1958* (New York: Praeger Publishers, 2000).
"Soviet Protests Canal Blockade," *New York Times*, November 5, 1956.
Spector, Leonard S., *Nuclear Ambitions: The Spread of Nuclear Weapons 1989–1990* (Boulder, CO: Westview Press, 1990).
———, *Nuclear Proliferation Today* (New York: Vintage, 1984).
———, *The Undeclared Bomb* (Cambridge, MA: Ballinger Publishing, 1988).
Spiegel, Steven L., *The Other Arab-Israeli Conflict: Making America's Middle East Policy from Truman to Reagan* (Chicago: University of Chicago Press, 1985).
Stein, Kenneth W., *Heroic Diplomacy* (New York: Routledge, 1999).
Steinbruner, John, "Consensual Security," *Bulletin of Atomic Scientists*, Vol. 64, No. 1 (March/April 2008), pp. 23–27.
Strange, Susan, "The Persistent Myth of Lost Hegemony," *International Organization*, Vol. 41, No. 4 (1987), pp. 551–573.
Strulak, Tadeusz, "The Nuclear Suppliers Group," *Nonproliferation Review*, Vol. 1, No. 1 (Fall 1993), pp. 2–10.
Stumpf, Waldo, "South Africa's Nuclear Weapons Program: From Deterrence to Dismantlement," *Arms Control Today*, Vol. 25, No. 10 (December 1995/January 1996), pp. 3–8.
Subrahmanyam, K., *India and the Nuclear Challenge* (New Delhi: South Asia Books, 1986).
———, *Nuclear Proliferation and International Security* (New Delhi: South Asia Books, 1985).
Sundarji, Lieutenant General K., ed., "Effects of Nuclear Asymmetry on Conventional Deterrence," *Combat Papers* (Mhow), No. 1 (April 1981).
———, ed., "Nuclear Weapons in the Third World Context," *Combat Papers* (Mhow), No. 2 (August 1981).
Ta Jen Liu, *U.S.-China Relations, 1784–1992* (New York: University Press of America, 2002).
Talmadge, Caitlin, "Deterring a Nuclear 9/11," *Washington Quarterly*, Vol. 30, No. 2 (Spring 2007), pp. 21–34.

Tannenwald, Nina, *The Nuclear Taboo: The United States and the Non-Use of Nuclear Weapons since 1945* (Cambridge: Cambridge University Press, 2006).
Telhami, Shibley, *The Stakes: America and the Middle East* (Boulder, CO: Westview Press, 2002).
Tellis, Ashley J., *India's Emerging Nuclear Posture* (Santa Monica, CA: RAND, 2001).
Tempest, Rone, "Dangerous Dynamic between China and India," *Los Angeles Times*, June 13, 1998.
Thayer, Bradley A., "The Causes of Nuclear Proliferation and the Nonproliferation Regime," *Security Studies*, Vol. 4, No. 3 (Spring 1995), pp. 463–519.
——, "The Risk of Nuclear Inadvertence: A Review Essay," *Security Studies*, Vol. 3, No. 3 (Spring 1994), pp. 428–493.
Thomas, Mary Ann, and Ramesh Santanam, "Government Agencies Investigated Missing Uranium, NUMEC," *Valley News Dispatch*, August 25, 2002.
Thucydides, *History of the Peloponnesian Wars*, Rex Warner, trans. (New York: Viking Press, 1954).
Trachtenberg, Marc, *A Constructed Peace: The Making of the European Settlement, 1945–1963* (Princeton, NJ: Princeton University Press, 1999).
Treaty on the Nonproliferation of Nuclear Weapons, 1968, available at www.fas.org/nuke/control/npt/text/npt2.htm.
Troen, Selwyn Ilan, and Moshe Shemesh, eds., *The Suez-Sinai Crisis 1956: Retrospective and Reappraisal* (New York: Columbia University Press, 1990).
Tsou, Tang, *China in Crisis: China's Policies in Asia and America's Alternatives* (Chicago: University of Chicago Press, 1969).
Tucker, Richard, *BTSCS: A Binary Time-Series—Cross-Section Data Analysis Utility*, version 4.0.4., Harvard University, 1999, available at www.fas.harvard.edu/~rtucker/programs/btscs/btscs.html.
Tufte, Edward, *Data Analysis for Politics and Policy* (Upper Saddle River, NJ: Prentice-Hall, 1974).
Ullman, Richard H., "The Covert French Connection," *Foreign Policy*, No. 75 (Summer 1989), pp. 3–33.
Ullom, Joel, "Enriched Uranium versus Plutonium: Proliferation Preferences in the Choice of Fissile Material," *Nonproliferation Review*, Vol. 2, No. 1 (Fall 1994), pp. 1–15.
URENCO, *URENCO Company History*, available at www.urenco.com/index.php?id=194&cid=305&gcid=317.
United States National Archives, Memorandum, George McGhee to Dean Rusk, September 13, 1961, Freedom of Information Act Files, India, USNA, College Park, MD.
Ur-Rehman, Shahid, *Long Road to Chagai* (Islamabad: Print Wise Production, 1999).
Van Evera, Steven, "The Cult of the Offensive and the Origins of the First World War," *International Security*, Vol. 9, No. 1 (Summer 1984), pp. 58–107.
——, "Primed for Peace: Europe after the Cold War," *International Security*, Vol. 15, No. 3 (Winter 1990/91), pp. 7–57.
Von Stein, Jana, "Do Treaties Constrain or Screen? Selection Bias and Treaty Compliance," *American Political Science Review*, Vol. 99, No. 4 (November 2005), pp. 611–623.
Walt, Stephen, "Alliance Formation and the Balance of Power," *International Security*, Vol. 9, No. 4 (Spring 1985), pp. 3–43.
——, *The Origins of Alliances* (Ithaca, NY: Cornell University Press, 1990).
Waltz, Kenneth, "More May Be Better," in Scott D. Sagan and Kenneth N. Waltz, eds., *The Spread of Nuclear Weapons: A Debate* (New York: W.W. Norton, 1995).
——, *Theory of International Politics* (New York: McGraw Hill, 1979).
Watson, Cynthia, *Argentine Nuclear Development: Capabilities and Implications*, Ph.D. dissertation, University of Notre Dame, 1984.

Way, Christopher, and Karthika Sasikumar, "Leaders and Laggards: When and Why do Countries Sign the NPT?" *REGIS Working Paper*, November, 17, 2004.

Weber, Steven, *Cooperation and Discord in US-Soviet Arms Control* (Princeton, NJ: Princeton University Press, 1991).

———, "Cooperation and Interdependence," *Daedalus*, Vol. 120, No. 2 (1991), pp. 183–201.

———, "Shaping the Postwar Balance of Power: Multilateralism in NATO," *International Organization*, Vol. 46, No. 3 (Summer 1992), pp. 633–680.

Weber, Steven, Naazneen Barma, Matthew Kroenig, and Ely Ratner, "How Globalization Went Bad," *Foreign Policy*, Vol. 86, No. 1 (January/February 2007), pp. 48–54.

Wehling, Fred, "Russian Nuclear and Missile Exports to Iran," *Nonproliferation Review*, Vol. 6, No. 2 (Winter 1999), pp. 134–143.

Weiner, Allen, "The Use of Force and Contemporary Security Threats: Old Medicine for New Ills?" *Stanford Law Review*, Vol. 59, No. 2 (2006), pp. 415–504.

Weiss, Leonard, "Atoms for Peace," *Bulletin of Atomic Scientists*, Vol. 59, No. 6 (November 2003).

Weissman, Steve, and Herbert Krosney, *The Islamic Bomb: The Nuclear Threat to Israel and the Middle East* (New York: New York Times Books, 1981).

Westad, Odd Arne, *The Global Cold War: Third World Interventions and the Making of Our Times* (Cambridge, U.K.: Cambridge University Press, 2007).

———, ed., *Brothers in Arms: The Rise and Fall of the Sino-Soviet Alliance, 1945–1963* (Washington, D.C.: Woodrow Wilson Center Press, 1998).

Whiting, Allen S., *The Chinese Calculus of Deterrence: India and Indochina* (Ann Arbor: University of Michigan Press, 1975).

Woodward, Bob, Robert G. Kaiser, David B. Ottaway, "U.S. Fears Bin Laden Made Nuclear Strides," *Washington Post*, December 4, 2001, A1.

World Nuclear Association, *Country Briefings: France*, available at www.world-nuclear.org/wgs/decom/projects/up1_print.htm.

———, "Nuclear Power Reactors," available at www.iaea.org/worldatom/rrdb/ and www.world-nuclear.org/wgs/decom/database/php/reactorsdb_index.php.

Wit, Joel S., Daniel B. Poneman, and Robert L. Gallucci, *Going Critical: The First North Korean Nuclear Crisis* (Washington, D.C.: Brookings, 2004).

Xia, Yafeng, *Negotiating with the Enemy: U.S.-China Talks during the Cold War, 1949–1972* (Bloomington: Indiana University Press, 2006).

Yoon Won-sup, "Park Sought to Develop Nuclear Weapons," *Korea Times*, January 15, 2008.

Young, Oran, "Regime Dynamics: The Rise and Fall of International Regimes," in Stephen Krasner, ed., *International Regimes* (Ithaca, NY: Cornell University Press, 1983).

Zeevy, Rechavam, "The Military Lessons of the Sinai Campaign: The Israeli Perspective," in Selwyn Ilan Troen and Moshe Shemesh, eds., *The Suez-Sinai Crisis 1956: Retrospective and Reappraisal* (New York: Columbia University Press, 1990).

Zhang, Shu, *Economic Cold War: America's Embargo against China and the Sino-Soviet Alliance, 1949–1963* (Stanford, CA: Stanford University Press, 2002).

Zissis, Carin, "The Fragile U.S.-South Korea Alliance: Backgrounder," *Council on Foreign Relations*, September 2006.

———, "The Six-Party Talks on North Korea's Nuclear Program," *Council on Foreign Relations*, June 2008.

Index

Pages numbers followed by *t* indicate tables.

Algeria, 70, 149, 161*t*, 199
Alliance hypothesis, 45–46, 170
Alliance structures, undermined by proliferation, 26–28, 34, 36*t*
Angola, 132
Arab-Israeli War, 30, 68, 89, 91
Argentina
   as capable nuclear supplier, 52*t*
   no assistance given, 39, 104–5, 110, 175, 202
      superpower dependence, 103–5
   no assistance received, 159, 203
Aswan Dam, 89
Atoms for Peace (U.S. program), 69, 157, 201
Australia, 203

Baker, A. J., 95
Balance of threat theory, 181–82
Ball, George, 31
Baradei, Mohamed El-, 11
Beg, Mirza Aslam, 138, 141–42
Belgium, 52*t*
Ben-Gurion, David, 24, 31, 72–73, 77, 85, 86–87
Bergmann, Ernst David, 68–70, 72, 75
Bhatia, Shyam, 139
Bhutto, Benazir, 137
Bhutto, Zulfikar Ali, 136
Bin Laden, Osama, 189
Blair, Dennis C., 8
Blanton, Shannon Lindsay, 183
Bourgès-Maunoury, Maurice, 71, 73, 100
Brazil, 5, 51, 52*t*, 161*t*, 198
Brookes, Peter, 22, 32–33
Brown, Charles R., 83
Bueno de Mesquita, Bruce, 53
Bulganin, Nikolai, 72
Bundy, McGeorge, 85
Bush, George H. W., 143
Bush, George W., 1, 22, 30, 141, 186

Canada, 200
Capable nuclear suppliers, 5, 51
   list of, 52*t*
Chamoun, Camille, 84

China
   as capable nuclear supplier, 52*t*
   changing policies of, 149–50, 176
   further proliferation, 31, 32
   no assistance given
      to Egypt, 159
      to Iraq, 202
      to North Korea, 202
   no assistance received, 132–34
   power-projecting status of, 15, 16, 112–14
   sensitive assistance given
      to Algeria, 161*t*, 199
      to Iran, 161*t*, 199
   sensitive assistance given, to Pakistan, 1, 38, 112–19, 155–56, 158, 161*t*, 175, 199
      alternative explanations, 117–19
      common enemies, 114–15
      power projection, 112–14
      superpower dependence, 116–17
   sensitive assistance received, from Soviet Union, 119–28, 156, 158, 161*t*, 175, 197
      alternative explanations, 126–28
      common enemies, 122–23
      nuclear proliferation, 156, 158
      power projection, 120–21
      superpower dependence, 124–26
   strategic costs of proliferation to, 18–19, 21–22, 24, 27
Climate change, 172, 185–86
Clinton, Bill, 32, 143
Coercive diplomacy, reduced by proliferation, 20–22, 34, 36*t*
Cohen, Avner, 19, 22, 71
Common enemies hypothesis, 37–38, 174
   China's assistance to Pakistan, 114–15
   France's assistance to Israel, 98–101
   India and assistance to Vietnam and Taiwan, 132–34
   Israel's assistance to South Africa, 131–32
   Pakistan's assistance to Iran, Libya, North Korea, 140–42
   Soviet Union's assistance to China, 122–23

Common enemies hypothesis *(continued)*
  statistical analysis of, 50, 53, 56t, 57–58t, 59, 62t, 63, 64
  strategic positions and, 88–89, 98–101
  U.S. and Israel's nuclear program, 88–89
Composite Index of National Capability (CINC), 51, 53
Conventional arms sales, 2, 182–85
  China and Pakistan, 115
  France and Algeria, 70–71
  France and Israel, 99, 108
  Soviet Union and China, 123
  U.S. and Israel, 78–79, 88–89
  U.S. and Pakistan, 143, 145
Corera, Gordon, 115, 142

De Gaulle, Charles, 26–27, 28, 75–76, 101–3, 109, 197–98
Diehl, Paul F., 53
Dimona nuclear site, 30, 74–79, 91, 158, 197–98
Dobrynin, Anatoly, 126

Eban, Abba, 87
Economic hypothesis, 4–5, 40–42
  China's assistance to Pakistan, 117–18
  France's assistance to Israel, 107–8
  nuclear proliferation and, 177, 187
  Pakistan's assistance to Iran, Libya, North Korea, 145–46
  Soviet Union's assistance to China, 126–27
  statistical analysis of, 54, 56t, 57–58t, 59–60
Edwards, John, 32
Egypt
  France and Israel's nuclear program, 70–73, 95–97, 98–100
  no assistance given to, 159
  sensitive assistance received, 161t, 198–99
  strategic costs of proliferation to, 19, 22, 30
  U.S. and Israel's nuclear program, 88–89
Eisenhower, Dwight D., 69, 72, 88, 201
Eisenhower Doctrine, 83–84
Eli, Paul, 99–100
Eshkol, Levi, 77, 87

Ferguson, Charles D., 172
Ford, Gerald F., 106
France
  as capable nuclear supplier, 5, 51, 52t
  collapse of empire, 93
  economic hypothesis, 41
  no assistance given
    to Australia, 203
    to South Korea, 202
  nonsensitive assistance given, 200
  NSG and, 61
  power-projecting status of, 15
  sensitive assistance given
    to Egypt, 161t, 198–99
    to Japan, 13, 161t, 198
    to Pakistan, 161t, 198
    to Taiwan, 105–6, 110, 161t, 198
  sensitive assistance given, to Israel, 1, 38, 67–68, 70–80, 161t, 174–75, 197–98
    alternative explanations, 106–9
    common enemies, 98–101
    nuclear proliferation and, 156, 158
    power projection, 92–98
    strategic positions, 91–103, 106, 109–10
    super power dependence, 101–3
  sensitive assistance received, 13, 201–2
Furhmann, Matthew, 157
Further proliferation, set off by proliferation, 31–33

Gaddafi, Muammar Al-, 136, 141
Gaddis, John Lewis, 128
Gartzke, Erik, 152, 164
Gavin, Francis, 30
Germany
  as capable nuclear supplier, 5, 51, 52t
  no assistance given
    to Iraq, 202
    to South Africa, 203
  nonsensitive assistance given, 200
  NSG and, 61
  sensitive assistance given
    to Brazil, 161t, 198
    to Iraq, 12–13
  sensitive assistance received, 13
Gilpatrick Committee, 21, 27, 85
Ginor, Isabella, 28, 91
Gleditsch, Kristian, 191
Global pandemics, 185–86
Goertz, Gary, 53
Goldschmidt, Bertrand, 72, 99
Goncharenko, Sergei, 127
Gowa, Joanne, 45
Gromyko, Andrei, 122
Ground invasion capability, as power-projecting qualifier, 14–15

Harmon, Avraham, 87
Harriman, Averell, 125
Hegemonic nonproliferation order. *See* Superpower dependence hypothesis
Hersh, Seymour, 75, 197–98
Highly enriched uranium (HEU), 11
Holloway, David, 30

INDEX

Hussein, king of Jordan, 83
Hussein, Saddam, 22
Hymans, Jacques, 153

Ideology, Soviet assistance to China and, 127–28
India
  as capable nuclear supplier, 52t
  as common enemy of China and Pakistan, 114–15
  further proliferation and, 31, 32
  no assistance given
    common enemies, 132–34
    power projection, 132–34
    superpower dependence, 133
  nonsensitive assistance received
    from Canada, 200
    from U.S., 200, 202
  strategic costs of proliferation to, 20, 23, 24, 25, 30
Indonesia, 31
Institutional hypothesis, 5, 42–44, 54
  China's assistance to Pakistan, 118–19
  France's assistance to Israel, 108–9
  nuclear proliferation and, 176–78
  Pakistan's assistance to Iran, Libya, North Korea, 146–47
  Soviet Union's assistance to China, 127
  statistical analysis of, 56t, 57–58t, 60
Inter-American Treaty of Reciprocal Assistance (Rio Pact), 104
International Atomic Energy Agency (IAEA), 11, 116
Iran
  further proliferation and, 32
  no assistance received
    from Argentina, 202
    from Germany, 202
  nonsensitive assistance received, 149, 201
  sensitive assistance received, from China, 161t, 199
  sensitive assistance received, from Pakistan, 2, 134–47, 158–59, 161t, 175–76, 199, 200
    alternative explanations, 145–47
    common enemies, 140–42
    power projection, 139–40
    superpower dependence, 142–45
  strategic costs of proliferation to, 19, 22, 26, 30, 33
Iraq
  no assistance received, 159, 202, 203
  nonsensitive assistance received, 200
  sensitive assistance needed, 155

  sensitive assistance received, 12–13, 161t, 198
  strategic costs of proliferation to, to U.S., 30
Ireland, 181
Israel
  as capable nuclear supplier, 52t
  further proliferation and, 31
  no assistance given, 130–32
  power projection, nonsensitive assistance to South Africa, 131–32, 175, 200
Israel, nuclear program development, 67–110
  background and overview
    France and, 70–74
    program origins, 68
    U.S. and, 69–70, 74–80
  French assistance to, 1, 38, 161t, 197–98
    alternative explanations, 106–9
    common enemies, 98–101
    power projection, 92–98
    superpower dependence, 101–3
  French strategic position, 91, 174–75
  nuclear proliferation and, 156, 158
  strategic costs of proliferation to, 19, 22–25, 28, 30
  U.S. strategic position, 80
    common enemies, 88–89
    power projection, 80–87
    superpower dependence, 89–91
Israeli Atomic Energy Commission (IAEC), 68
Italy, 52t, 161t, 198, 203

Jabko, Nicholas, 41, 107
Japan
  as capable nuclear supplier, 5, 39, 52t
  further proliferation and, 31, 32
  nonsensitive assistance received, 200
  sensitive assistance received, 13, 161t, 198, 200
Jo, Dong-Joon, 152, 164
Johnson, Lyndon B., 21, 78, 85, 86, 87, 126
Johnston, Alastair Iain, 115
Jones, Lewis G., 87

Karamat, Jehangir, 140
Karnad, Bahrat, 133
Kennedy, John F.
  China's nuclear program, 18–19, 21, 125
  Israel's nuclear program, 23–24, 31, 76–79, 85, 86–87
Kent, Sherman, 85, 86
Khan, A. Q., 1–2, 134–39, 142–44, 146–48, 159, 201
Khan, Gulam Ishaq, 138, 143
Khan, Hendrina, 138
Khrushchev, Nikita, 125, 128
Kibaroglu, Mustafa, 19, 33

Kim Il Sung, 137, 139
Kissinger, Henry, 24, 79–80
Klein, James, 53
Komer, Robert, 85

Lavoy, Peter, 157
Lebanon, 83–84, 95
Leghari, Farooq, 139
Lewis, John, 122
Li Fuchan, 121
Libya, 105, 110
  no assistance received, 39, 110, 175, 202
  nonsensitive assistance received, 201
  sensitive assistance received, from Pakistan, 2, 134–47, 158–59, 161t, 175–76, 200, 201
    alternative explanations, 145–47
    common enemies, 140–42
    power projection, 139–40
    superpower dependence, 142–45
    strategic costs of proliferation to, 30
Louis, Victor, 124
Lucet, Charles, 77

Manhattan Project, 13, 203
Mao Zedong, 122, 128
Martin, Michel, 94
McCloy, John, 27
McCone, John, 76, 77
McGhee, George, 31
Mearsheimer, John, 15
Medeiros, Evan, 199
Meir, Golda, 73
Middle East, U.S. and Israel's nuclear program, 80–91
Military intervention capabilities, deterred by proliferation, 17–20, 34, 36t
Mollet, Guy, 95, 99, 100, 197–98
Montgomery, Alex, 154
Mozambique, 132
Mullins, Martin, 104
Musharaff, Pervez, 138–39, 144, 199

Nasser, Gamal Abdel, 30, 70–73, 85, 89, 95, 98–100
National Intelligence Estimate (NIE), 1963 U.S., 18, 21, 23
Netherlands, 52t, 201
New Zealand, 181
Nixon, Richard M., 79–80, 105
Nonpower-projecting states, 173–77
  defined, 14
  differential effects of proliferation, 3–4, 34–36, 36t, 180–81

Nonsensitive nuclear assistance, nuclear proliferation and, 156–57, 162–63, 169–71. *See also specific countries*
North Korea
  as capable nuclear supplier, 52t
  further proliferation and, 32
  no assistance received, 202
  nonsensitive assistance given, 201
  nonsensitive assistance received, 201
  reactions to nuclear test, 1
  sensitive assistance received, from Pakistan, 2, 134–47, 158–59, 161t, 175–76, 200, 201
    alternative explanations, 145–47
    common enemies, 140–42
    power projection, 139–40
    superpower dependence, 142–45
    strategic costs of proliferation to, 19, 24–27, 30–31, 186
North Korea Potential for Nuclear Weapons Development (CIA assessment), 19
Norway, 52t, 203
Nuclear domino effect, 31
Nuclear Nonproliferation Treaty (NPT), 38, 43, 170, 176–78
  China and, 116, 118, 149
  nuclear proliferation and, 169–71
  statistical analysis of, 56t, 57–58t, 60, 62t, 64–65, 192
Nuclear proliferation, assistance and, 1, 151–52
  case studies, 158–59
  conclusion, 171–72
  data, 159–63, 195–97
  data analysis, 163–71
  demand-side approach, 152–53
  nonsensitive assistance and, 156–57
  sensitive nuclear assistance and, 154–56
  supply-side approach, 3, 6, 153–54, 178–79
Nuclear proliferation, differential effects of, 16–34, 179–81
  deters military intervention, 17–20, 34, 36t
  dissipates strategic attention, 28–31, 34, 36t
  reduces effectiveness of coercive diplomacy, 20–22, 34, 36t
  sets off further nuclear proliferation, 31–33, 34, 36t
  triggers regional instability, 22–26, 34, 36t
  undermines alliance structures, 26–28, 34, 36t
Nuclear Proliferation Prevention Act, 116
Nuclear proliferation, prevention of, 173–90
  balance of threat theory, 181–82
  conventional arms sales and, 182–85
  strategic costs and, 173–78
  supply-side approach, 178–79

transnational security threats and, 185–86
U.S. foreign policy implications, 186–90
Nuclear proliferation, strategic theory of, 10, 16–40
　alternative explanations, 40–46
　elements of argument, 10–16
　hypotheses, 4, 36–40
　nonpower-projecting states less threatened by strategic costs, 3–4, 34–36, 36t
　power-projecting states more threatened by strategic costs, 3, 16–34, 36t
　research design and case selection, 46–49, 47t
Nuclear Security Council (NSC), 38, 43
Nuclear Suppliers Group (NSG)
　China and, 117, 118, 149
　statistical analysis of, 56t, 57–58t, 60–61
Nuclear terrorism, 35, 189–90
Nuclear weapons possession hypothesis, 5, 44–45, 146–47
　China's assistance to Pakistan, 119
　France's assistance to Israel, 109
　Pakistan's assistance to Iran, Libya, and North Korea, 147
　Soviet Union's assistance to China, 127
　statistical analysis of, 56t, 57–58t, 59

Oneal, John R., 191
Optimism/pessimism debate, 6–7, 179–81
Origins of Alliances, The (Walt), 181–82

Pakistan
　as capable nuclear supplier, 5, 51, 52t
　further proliferation and, 31, 32
　no assistance given, 203
　nonsensitive assistance given, 201
　nonsensitive assistance received, 201
　sensitive assistance given, to Iran, Libya, North Korea, 2, 134–47, 158–59, 161t, 175–76, 199, 200
　　A. Q. Khan's role in, 134–39
　　alternative explanations, 145–47
　　common enemies, 140–42
　　power projection, 139–40
　　superpower dependence, 142–45
　sensitive assistance received, from China, 1, 38, 112–19, 155–56, 158, 161t, 175, 199
　　alternative explanations, 117–19
　　common enemies, 114–15
　　power projection, 112–14
　　superpower dependence, 116–17
　sensitive assistance received, from France, 161t, 198
　strategic costs of proliferation to, 20, 23, 25, 30

Pape, Robert, 15
Paul, T. V., 42, 115
People's Republic of China. *See* China
Peres, Simon, 68, 72–73, 87, 98–100, 108
Perrin, Jean-Francis, 103
Pilat, Joseph, 44
Pineau, Christian, 70, 73, 99
Pollack, Jonathan, 44
Posen, Barry, 22
Potter, William, 40, 42
Power projection hypothesis, 36–37, 47t, 173–77
　China's assistance to Pakistan, 112–14
　France's assistance to Israel, 92–98
　Israel's nonassistance to South Africa, 131–32, 175, 200
　India's nonassistance to Vietnam and Taiwan, 132–34
　Pakistan's assistance to Iran, Libya, North Korea, 139–40
　power projection defined, 14–16
　Soviet Union's assistance to China, 120–21
　statistical analysis of, 50, 55, 56t, 57–58t, 62–63, 62t, 63f
　strategic positions and, 80–87, 91–98
　U.S. and Israel's nuclear program, 80–87
Power-projecting states, differential effects of proliferation, 3, 16–34, 36t, 180–81
　alliance structure undermined, 26–28
　coercive diplomacy reduced, 20–22
　counterarguments, 33–34
　further proliferation and, 31–33
　military intervention deterred, 17–20
　regional instability triggered, 22–26
　strategic attention dissipated, 29–31
Proliferation optimists/pessimists. *See* Optimism/pessimism debate

Quester, George, 44

Rare Events Logistic Regression (ReLogit), 54–55
Reed, Thomas, 115
Regime type, 192
Regional instability, triggered by proliferation, 22–26, 34, 36t
Reid, Ogden, 87
Remez, Gideon, 28, 91
Richelson, Jeffrey, 78
Rio Pact, 104
Rostow, Walt, 125
Rowen, Henry, 24, 143
Rusk, Dean, 76, 84, 86–87, 126
Russett, Bruce, 191

232 INDEX

Russia. *See also* Soviet Union
  as capable nuclear supplier, 52t
  changing policies, 148–50, 176
  economic hypothesis, 41
  nonsensitive assistance to Iran, 201

Sagan, Scott D., 7, 152–53, 180
Saint-Gobain Techniques Nouvelles (SGN), 74
Sajjad, Wasim, 139
Samore, Gary, 139
Saudi Arabia, 159
Scheinman, Lawrence, 102
Schelling, Thomas, 27
Seaborg, Glen, 87
Sensitive nuclear assistance. *See also specific countries*
  cases of assistance, 161t, 197–200
  cases of no assistance, 202–3
  cases of nonsensitive assistance, 200–202
  excluded activities, 12–14
  frequency of, 129–30, 148–50
  further research directions, 178–79
  included activities, 2, 10–12, 14
  nuclear proliferation and, 154–56, 161–64, 161t, 165–68t, 169
  as permanent solution to temporary problem, 189
  states never providing, 104
Sensitive nuclear assistance hypotheses.
  See Alliance hypothesis; Common enemies hypothesis; Economic hypothesis; Institutional hypothesis; Power projection hypothesis; Superpower dependence hypothesis
Sensitive nuclear assistance, statistical analysis of, 50–66
  conclusion, 65–66
  data analysis, 54–65
  data set variables, 50–54, 191–95
Serbia, 52t
Sharif, Nawaz, 143–44
Singh, Sonali, 163–64, 191, 193
Skinner, Selby, 32
Smuggling, of nuclear technology, 13–14, 146–48
Solingen, Etel, 42, 153
South Africa
  as capable nuclear supplier, 52t
  further proliferation and, 32
  no assistance received, 203
  nonsensitive assistance from Israel, 175, 200
    common enemies, 132
    power projection, 130
    superpower dependence, 131–32
  strategic costs of proliferation to, 25, 30

South Korea
  further proliferation and, 31, 32
  no assistance to, 159, 202
  strategic costs of proliferation to, 25–28, 30–31
Soviet Union. *See also* Russia
  as capable nuclear supplier, 52t
  Eisenhower Doctrine and limit of influence, 83–84
  Israel's nuclear program development, 88, 90–91, 110, 132
  no assistance given, 159
  nonsensitive assistance given, 201
  sensitive assistance given, to China, 119–28, 161t, 175, 197
    alternative explanations, 126–28
    common enemies, 122–23
    nuclear proliferation and, 156, 158
    power projection, 120–21
    superpower dependence, 124–26
  strategic costs of proliferation, 22, 25, 28, 30
  Suez Canal Crisis and, 72–73
  U.S. and Israel's nuclear program, 88
  U.S. power projection and, 85–86
Steinbruner, John, 185
Stillman, Danny, 115
Strategic attention, dissipated by proliferation, 28–31
Strategic quid pro quo, Pakistan and, 147
Strategic Theory of Nuclear Proliferation.
  See Nuclear proliferation, strategic theory of
Strauss, Levi, 70
Suez Canal Crisis, 70–73, 88–89, 95–97, 98–100
Sundarji, K., 20
Superpower dependence hypothesis, 38–39, 174
  Argentina and, 103–5
  China's assistance to Pakistan, 116–17
  France and Israel's nuclear program, 101–3
  India and Vietnam and Taiwan, 132–34
  Israel and South Africa, 131–32, 175, 200
  Pakistan's assistance to Iran, Libya, North Korea, 142–45
  Soviet Union's assistance to China, 124–26
  statistical analysis of, 50
  strategic positions and, 89–91, 101–6
  Taiwan and, 105–6, 110
  U.S. and Israel's nuclear program, 89–91
Superpower pact variable, in statistical analysis, 53, 56t, 57–58t, 59, 62t, 63, 64
Superpower vote variable, in statistical analysis, 53, 56t, 57–58t, 59, 61, 62t
Supply-side approach, to causes of proliferation, 3, 153–54, 178–79
Syria, 159, 201, 203

Taiwan, 28, 31, 32, 159
　no assistance received, from India
　　common enemies, 132–34
　　power projection, 132–34
　　superpower dependence, 133
　sensitive assistance received, from France, 161*t*, 198
　　superpower dependence, 105–6, 110
Taiwan Relations Act (TRA), 105
Talbott, Strobe, 144
Tamir, Avraham, 132
Tarar, Muhammad Rafiq, 139
Tellis, Ashley, 20
Terrorism, threats of, 35, 185–86, 189–90
Transnational security threats, nuclear proliferation and, 185–86
Truman, Harry S., 105
Turkey, 19, 22, 26, 33

United Arab Republic, 31, 84
United Kingdom, 13, 52*t*, 202, 203
United Nations General Assembly (UNGA), voting behavior, 53–54
United States
　as capable nuclear supplier, 5, 51, 52*t*
　China and, 116–17, 119, 122–23, 124–26
　foreign policy implications, 186–90
　Israel's nuclear program development, 67–70, 74–80, 156, 174–75
　　alternative explanations, 106–9
　　common enemies, 88–89
　　power projection, 80–87
　　strategic positions, 80–91, 109–10
　　superpower dependence, 89–91
　no assistance given, 203
　nonsensitive assistance given, 13, 200–202
　NSG and, 61
　Pakistan's assistance to Iran, Libya, North Korea, 140–45
　strategic costs to U.S. of proliferation, 29–30
　　to China, 18–19, 21–22
　　to India, 24
　　to Iran, 30
　　to Iraq, 30
　　to Israel, 22, 23–24, 28
　　to Libya, 30
　　to North Korea, 24, 30

Vietnam
　no assistance received, from India
　　common enemies, 133, 175
　　power projection, 16
　　superpower dependence, 132–34
Vitro International, 202

Walt, Stephen, 181–82
Waltz, Kenneth, 7, 179–80
Way, Christopher R., 163–64, 191, 193
Weber, Steven, 41, 107
Weiner, Allen, 185
Weiss, Leonard, 157

Xue Litai, 122

Yom Kippur War, 1973, 24, 28
Yugoslavia, 52*t*, 202, 203

Zangger Committee, 38